SpringerBriefs in Law

SpringerBriefs present concise summaries of cutting-edge research and practical applications across a wide spectrum of fields. Featuring compact volumes of 50 to 125 pages, the series covers a range of content from professional to academic. Typical topics might include:

- A timely report of state-of-the art analytical techniques
- A bridge between new research results, as published in journal articles, and a contextual literature review
- A snapshot of a hot or emerging topic
- A presentation of core concepts that students must understand in order to make independent contributions

SpringerBriefs in Law showcase emerging theory, empirical research, and practical application in Law from a global author community.

SpringerBriefs are characterized by fast, global electronic dissemination, standard publishing contracts, standardized manuscript preparation and formatting guidelines, and expedited production schedules.

More information about this series at http://www.springer.com/series/10164

Dana Beldiman · Constantin Blanke-Roeser

An International Perspective on Design Protection of Visible Spare Parts

Springer

Dana Beldiman
Academic Director and Founder, Center for Transnational Intellectual Property
Bucerius Law School
Hamburg
Germany

and

UC Hastings College of the Law
San Francisco
USA

Constantin Blanke-Roeser
Researcher, Center for Transnational Intellectual Property
Bucerius Law School
Hamburg
Germany

ISSN 2192-855X ISSN 2192-8568 (electronic)
SpringerBriefs in Law
ISBN 978-3-319-54059-7 ISBN 978-3-319-54060-3 (eBook)
DOI 10.1007/978-3-319-54060-3

Library of Congress Control Number: 2017934314

Printed on acid-free paper

This Springer imprint is published by Springer Nature
The registered company is Springer International Publishing AG
The registered company address is: Gewerbestrasse 11, 6330 Cham, Switzerland

Preface and Acknowledgements

A few words about the context and the genesis of this publication.

Policymakers have long struggled with the question whether design protection for spare parts of a complex product extends to the aftermarket (use of parts for repair purposes), or whether that market should remain open to competition. The decision primarily impacts the automotive industry and involves control over the lucrative and highly competitive market for automotive visible spare parts. From a legal perspective the question hinges on a provision of industrial design law, known as the repair clause. This clause is intended to exempt from design protection visible spare parts, to the extent they are used for repair purposes and in order to restore the original appearance of a complex product.

In the EU the debate whether a repair clause should be included into the Design Directive (98/71/EC), has been ongoing for years. Adoption of the clause in the Directive would result in a binding obligation on member states and has been so far prevented by some member states, including Germany. Positions on the issue are divided. Supporters include spare parts manufacturers, insurance and consumer organizations, whereas opponents are manufacturers of complex products, primarily the automotive industry.

Against this background, the Center for Transnational IP, Media and Technology Law and Policy of Bucerius Law School, Hamburg, Germany (Bucerius IP Center) held on 28 October 2015 a full-day Expert Roundtable Discussion on the above topic. The Roundtable assembled leading experts from academia, industry, policy and law for a discussion aimed at identifying approaches which might lead to possible compromise positions among the stakeholders in the debate.

The following experts participated in this Roundtable:

- Marianne Grabrucker (President of the Circle of European Trademark Judges; former Presiding Judge at the Federal Patent Court of Germany)
- Professor Dr. Philipp Fabbio (Università Mediterranea di Reggio Calabria, Italy)
- Professor Dr. h.c. Annette Kur (Max Planck Institute for Innovation and Competition, Munich)

- Professor Dr. Dana Beldiman (Academic Director of Bucerius IP Center)
- Dr. Jutta Figge (Federal German Ministry of Justice and Consumer Protection)
- Dr. Peter Stiefel (Markenrechtsschutz, Daimler AG)
- Robert Sterner (Leiter Marken- und Produktschutz, Audi AG)
- Dr. Ingo Gehring (IP Legal, Siemens AG)
- Dr. Gerhard Riehle, LL.M. (Former CEO, The European Campaign for the Freedom of the Automotive Parts and Repair Market, ECAR)
- Johannes Hübner (Automobilclub von Deutschland e.V., AvD) *(unable to attend; provided information in writing)*

The meeting was held at the Bucerius Law School in Hamburg and was moderated by Marianne Grabrucker, President of the Circle of European Trademark Judges and former Presiding Judge at the Federal Patent Court of Germany.

The present publication was inspired by the intellectually stimulating and spirited discussion among the Roundtable participants. However, the facts and opinions stated herein are solely those of the authors and do not necessarily reflect the opinions of any of the Roundtable participants. The publication is also not to be understood as the outcome of the meeting or as a common understanding reached among the participants to the meeting. In fact, because of the multilayered nature of the issues that prompted further research and because of legal developments which occurred in the interim, this publication ended up covering many more and different points than the Roundtable. The authors have sought to consider the issues from multiple different vantage points and to provide a balanced, impartial view of IP-related issues surrounding the spare parts debate.

Our thanks go to the participants in the Roundtable Discussion for having pointed to this fascinating area of the law and for their contributions to the Roundtable discussion. In particular, we appreciate the efforts of Marianne Grabrucker, who has inspired the project and was able to lead the debate with extraordinary thoughtfulness and skill, thus ensuring the meeting's constructive and creative atmosphere. Ms. Grabrucker also provided important support in the preparation of the Roundtable and development of the main discussion points. Further, our thanks go to the ZEIT-Stiftung Ebelin und Gerd Bucerius for its generous financial support. Finally, we express our appreciation to Karsten Windler, the Executive Director of the Bucerius IP Center, and Vera Kortfunke, student assistant, for their organizational support, as well as the management of Bucerius Law School, represented by Meinhard Weizman, CEO, Dr. Hariolf Wenzler, former CEO, and by Benedikt Landgrebe, deputy CEO of Bucerius Law School for the ongoing support and encouragement for the activities of the IP Center. Moreover, we thank Prof. Fabiano Teodoro de Rezende Lara, Universidade Federal de Minas Gerais, Brazil, for his contribution to the section on the legal situation in Brazil.

Finally, we would like to thank the following for valuable comments to the draft: Alberto Camusso, Marianne Grabrucker, Prof. Dr. Philipp Fabbio, Dr. Peter Stiefel and Dr. Gerhard Riehle, LL.M.

We hope the material in this book will contribute to further scholarly legal debate on the topic.

Hamburg, Germany
November 2016

Dana Beldiman
Constantin Blanke-Roeser

Contents

Abbreviations

ABPA	Automotive Body Parts Association
AIPLA	American Intellectual Property Law Association
ANFAPE	Associaçao Nacional dos Fabricantes de Autopeças
APAA	The Asian Patent Attorneys Association
Art(s).	Article(s)
Berkeley Technology L J	Berkeley Technology Law Journal
BGH	Bundesgerichtshof
BR-Drucks.	Bundesrat-Drucksache
BT-Drucks.	Bundestag-Drucksache
CADE	Conselho Administrativo de Defesa Econômica
CAPA	Certified Automobile Parts Association
Cardozo Arts & Entertainment L J	Cardozo Arts & Entertainment Law Journal
CEAHR	European confederation for watch repairer associations
cf.	Conferre (compare)
DePaul J Art, Tech & Intell Prop L	The DePaul Journal of Art, Technology & Intellectual Property Law
e.g.	Exempli gratia (for example)
ECTA	European Communities Trademark Association
ECWVTA	European Community Whole Vehicle Type Approval
ed(s)	Editor(s)
edn.	Edition
et al.	Et alii (and others)
et seq.	Et sequentes (and the following)
etc.	Et cetera (and so forth)
EWS	Europäisches Wirtschafts- und Steuerrecht

FIGIEFA	International Federation of Automotive Aftermarket Distributors
FS	Festschrift
GDP	Gross domestic product
George Mason L R	George Mason Law Review
Gonzaga L Rev	Gonzaga Law Review
GRUR	Gewerblicher Rechtsschutz und Urheberrecht
GRUR Ausl	Gewerblicher Rechtsschutz und Urheberrecht, Auslands- und internationaler Teil (1952-1966, ab 1967 GRUR Int)
GRUR-Beil	Gewerblicher Rechtsschutz und Urheberrecht – Beilage
GRUR-Int	Gewerblicher Rechtsschutz und Urheberrecht Internationaler Teil
GRUR-Prax.	Gewerblicher Rechtsschutz und Urheberrecht Praktischer Teil
GRUR-RS	Gewerblicher Rechtsschutz und Urheberrecht
Harvard J L & Tech	Harvard Journal of Law & Technology
Hastings L J	Hastings Law Journal
i.e.	Id est (that is)
IAM	Independent Aftermarket
IIC	International Review of Intellectual Property and Competition Law
Int J of Production	International Journal of Production
ISRN	International Scholarly Research Network
ISRN Mechanical Engineering	International Scholarly Research Notes Mechanical Engineering
J Manuf Sci Eng	Journal of Manufacturing Science and Engineering
J Pat & Trademark Off Soc'y	Journal of the Patent and Trademark Office Society
JICE	Journal of International Commerce and Economics
JIPITEC	Journal of Intellectual Property, Information Technology and E-Commerce Law
K&R	Kommunikation & Recht
Ky. L J	Kentucky Law Journal
Ltd.	Limited
Ltda.	Limitada
MarkenR	Markenrecht
Maryland L R	Maryland Law Review
Mitt	Mitteilungen der deutschen Patentanwälte
MMR	MultiMedia und Recht
NAIC	National Association of Insurance Commissioners

NHTSA	National Highway Transportation Safety Administration
NJW	Neue Juristische Wochenschrift
nr.	Number
NYU L R	The New York University Law Review
OEM	Original Equipment Manufacturers
OES	Original Equipment Suppliers
p(p).	Page(s)
para	Paragraph
R&D	Research and Development
Rs.	Rechtssache
Santa Clara High Technology L J	Santa Clara High Technology Law Journal
SDE	Secretaría de Economía
sec.	Section
SME	Small- and medium-sized enterprises
Southern University L R	Southern University Law Review
U Balt L Rev	The University of Baltimore Law Review
UNECE	United Nations Economic Commission for Europe
USPTO	United States Patent and Trademark Office
v.	Versus
WIPO	World Intellectual Property Organization
WTO	World Trade Organization

Chapter 1
Introduction

Use of complex manufactured products commonly results in wear or breakage of some of their component parts. Auto parts, razor blades, printer cartridges and others all need to be replaced periodically. For consumers, replacement of a part is usually more advantageous than acquisition of a new product.[1] This makes production of replacement parts (spare parts) a lucrative business and gives rise to competition in the market. Producers of complex products feel entitled to supply spare parts on a continued basis by virtue of having created the product which originally incorporated the part in the first place.[2] New companies, which specialize solely in the production of spare parts seek market entry in order to take advantage of the opportunities offered by spare parts markets.[3]

In the early days of *en masse* industrial production of durable goods, replacement parts were usually produced by the original manufacturer of the complex product. Having developed the product, including its parts, the original manufacturer possessed the requisite know-how, the production capabilities, as well as the relationship with the purchaser. As a result, a product's original manufacturer was also typically the source of all replacement parts compatible with the original complex products. An automotive manufacturer would, for instance, routinely

[1]Except for some relatively low cost common usage items, for which part replacement is overall not economically justified.

[2]Original manufacturers of complex products use a variety of means to maintain control of the aftermarket, such as technical means (e.g. computer codes for printer cartridges), advertising, sales and marketing techniques, part numbering systems, contractual arrangements with the distribution chain.

[3]In some industries producers of original products resort to business models of setting artificially low prices for original products, while pricing component parts high. This strategy works particularly well in products which require regular "refills" such as printer cartridges (see e.g. Berns 2013, note 9, with reference to Gerber 2005), electric toothbrushes and coffee makers. The success of this strategy was proven early on by Standard Oil's introduction of a safe and long burning kerosene lamp, the *Mei-Foo* lamp in China. The lamp itself was sold for almost nothing and even often given away for free with the purchase of kerosene. The profits came from the ensuing sales of kerosene, which opened up the Chinese market for Standard Oil; see Berns 2013, p. 1.

D. Beldiman and C. Blanke-Roeser, *An International Perspective on Design Protection of Visible Spare Parts*, SpringerBriefs in Law,
DOI 10.1007/978-3-319-54060-3_1

produce the replacement parts for its vehicles. This paradigm changed in the 1970–1980s, as a result of several developments, including less costly manufacturing technologies, growth of overseas production and the emergence of container shipping. The overall cost of replicating, manufacturing and distributing replacement parts dropped dramatically and gave rise to opportunities for new players to enter the market.[4] This in turn prompted original manufacturers to seek to protect their market positions with respect to spare parts, among other things, by use of IP protection to prevent imitation of their spare parts by unauthorized parties. Spare parts manufacturers responded by lobbying for liberalization of the spare parts market, specifically by seeking legislation to eliminate IP protection. These efforts have been successful in some jurisdictions, but not in others. The dispute is still ongoing.[5]

From a legal perspective, the spare parts debate revolves to a large extent around IP laws. These laws were conceived to incentivize innovation and protect original manufacturers against imitation and counterfeits. The following discussion will focus on the subset of IP laws referred to as industrial design or, simply, design law, which applies *inter alia* to spare parts. Design law covers a product's shape, design or other outer appearance,[6] by providing protection to component parts that are integral to its appearance and are visible to the user during normal use.[7] Because, in general, replacements for visible parts must constitute an exact match to the original parts in order not to alter the product's overall appearance, the spare parts at issue are also referred to as "must match" parts.

---

[4]These new entrants were perceived, to some extent, as unfairly competing by free-riding on the primary producers' efforts. Increased assertion of trademarks by primary producers led to a terminological division of the market into "original" and "non-original" equipment (see Berns 2013, p. 8).

[5]Contributing to the lack of progress is the absence of recent comprehensive data that would verify the stakeholders' contentions and allow a more accurate assessment of the situation. The European Commission's Proposal for a Directive of the European Parliament and of the Council amending Directive 98/71/EC on the legal protection of designs (COM(2004) 582 final – 2004/0203 (COD)), had gathered a certain amount of data upon which it based its findings, but by its own admission, these data were often not sufficient to be conclusive. Other position papers following it (e.g. Europe Economics The Economic Review of Industrial Rights in Europe – Final Report, MARKT/2013/064//D2ST/OP, January 2015, commissioned by the European Commission), added some new data, but partly relied on prior data, as did various position papers following it. See ECTA, Design Committee, Position Paper Designs and Spare Parts, 2016. The French Competition Authority's report of 2012 (Autorité de la concurrence, Avis n° 12-A-21 du 8 octobre 2012 relatif au fonctionnement concurrentiel des secteurs de la réparation et de l'entretien de véhicules et de la fabrication et de la distribution de pièces de rechange) has assembled information up the years 2010–2011. Yet other data sets are promoted by lobby associations. Overall, however, given the speed of technological development, data which are several years old, must be considered with caution.

[6]See infra chapter 3.

[7]Cf. Kur 2016, p. 21: For instance, parts of cars which are not visible, are brakes and couplings. In contrast, the visible parts are mainly those parts that are affected in case of a crash ("crash parts"). For component parts, see in particular infra section 3.3.

To what extent must match parts should be protected by design law has been debated for years, but the issue remains unresolved. More recently, the conflict has heated up, evidenced by increased legislative and judicial action relating to spare parts, primarily in the automotive field.

This book will discuss various aspects surrounding this conflict and will suggest considerations for compromise. The first chapter provides an overview over some of the business aspects in order to illuminate the source of tension in the spare parts debate. It is followed by a detailed examination of the provisions of the applicable design laws on a comparative basis in the EU and the U.S., with reference to a few other relevant jurisdictions. Next, the discussion focuses on legislative efforts to enact a "repair clause", i.e. legislation to exempt from design protection those visible spare parts which are used for repairs to restore the original appearance of the product, followed by judicial developments on the topic. Finally, we will list and assess a few considerations that may contribute to a possible compromise solution and conclude with a look at how future technological developments might impact the spare parts debate.

References

Autorité de la concurrence (2012) Avis n° 12-A-21 du 8 octobre 2012 relatif au fonctionnement concurrentiel des secteurs de la réparation et de l'entretien de véhicules et de la fabrication et de la distribution de pièces de rechange (The French Competition Authority's report of 2012). http://www.autoritedelaconcurrence.fr/pdf/avis/12a21.pdf

Berns E (2013) Marktmissbrauch auf Ersatzteilmärkten im deutschen, europäischen und US-amerikanischen Markt [Diss.]. EUL Verlag, Lohmar

European Commission (2004) Proposal for a Directive of the European Parliament and of the Council amending Directive 98/71/EC on the legal protection of designs (COM(2004) 582 final—2004/0203 (COD)). http://www.eumonitor.eu/9353000/1/j9vvik7m1c3gyxp/vi7jgsymsdz3

Europe Economics (2015) The Economic Review of Industrial Rights in Europe—Final Report, MARKT/2013/064//D2ST/OP, Jan 2015, commissioned by the European Commission. http://ec.europa.eu/growth/tools-databases/newsroom/cf/itemdetail.cfm?item_id=8844

European Communities Trademark Association (ECTA), Design Committee (2016) Position Paper Designs and Spare Parts. http://www.ecta.org/IMG/pdf/ECTA_-_Position_Paper_-_Spare_Parts_(003).pdf

Gerber T (2005) Canon greift zur Chipverdongelung bei Druckerpatronen. http://www.heise.de/newsticker/meldung/Canon-greift-zur-Chipverdongelung-bei-Druckerpatronen-124095.html

Kur A (2016) Ersatzteilfreiheit zwischen Marken- und Designrecht. GRUR 20–30

Chapter 2
Business Aspects of the Spare Parts Industry

2.1 Factors Affecting the Existence of a Spare Parts Market

Because production of compatible component parts can provide a steady source of revenue, it constitutes an attractive business model for manufacturers of both original products and of spare parts. For original product manufacturers, revenue from spare parts represents, in a favorable scenario, a source of profit and in a less favorable one, a hedging mechanism which allows them to subsidize the cost of the original manufacture in the event of economic downturns. For new entrants, the spare parts market is attractive because of relatively low development costs and prospects of a high profit margin.

In some industries, such as the automotive industry, robust spare part markets have developed, but that is not necessarily true for all industries. Whether a market exists is largely a function of demand, which, in turn, depends on a number of dynamics. Specifically, in a spare part market, demand is stimulated by repeat purchases of replacements for broken parts, which tend to continue throughout a product's entire lifecycle.[1] Demand persists even when the need for the original products decreases, be it for reasons of market saturation, or economic downturn. In a down economy, consumers tend to rely on repairing and replacing parts on existing goods, rather than on more costly new acquisitions. Finally, demand is a function of the number of products in circulation. A high level of consumption of goods increases the total number of goods sold, and with it the need for replacement parts.[2]

[1]See Berns 2013, p. 1; Riehle, FS Möschel 2011, p. 1076.
[2]See Berns 2013, p. 3; Riehle, FS Möschel 2011, p. 1076.

D. Beldiman and C. Blanke-Roeser, *An International Perspective on Design Protection of Visible Spare Parts*, SpringerBriefs in Law,
DOI 10.1007/978-3-319-54060-3_2

Other dynamics however, operate to counteract demand for spare parts. The acceleration of product lifecycles diminishes the demand for spare parts. Its two primary causes are technological and aesthetic obsolescence. The speed of technological development results in products becoming rapidly obsolete, due, for instance, to incompatibility with the latest state of technology. Consider the massive amounts of electronic equipment, which in theory is still functional, but in practice has become unusable. Aesthetic obsolescence occurs due to an accelerated succession in product styles, often as the result of innovative marketing strategies, which accelerate the consumption cycle.[3] Therefore, many modern products, primarily consumer products in a lower cost range, such as, for instance, coffee makers, mostly need no replacement parts and a market for such parts is unlikely to develop.

In higher cost products, additional factors may diminish demand. For instance, compliance with safety regulations, particularly in potentially hazardous products may raise production costs to a point where producing spare parts is unattractive. A computed tomography (CT) scanning device, for example, is potentially dangerous to both patients and third parties. The applicable safety regulations for, say the machine's outer hull, increase production costs to a point that the aftermarket for this product becomes unattractive. The risk of liability for defective products may have a similar effect, particularly in jurisdictions with consumer-friendly product liability laws, such as the U.S., where damage awards can be substantial. Technological complexity may also make the replacement part business unattractive. Consider a wind turbine, whose design protectable parts are its outer shell and parts of its tower. Production of such parts would require a high degree of specialization and make an aftermarket unprofitable. Finally, a spare parts market may not develop where policies of certain purchasers, such as government entities' procurement departments, require spare that parts be supplied by original manufacturers only, for instance, in products of critical importance, where product failure can simply not be allowed.

Combined, these circumstances have apparently favored the emergence of a competitive spare parts market primarily in the automotive industry and made it the target of regulation. Whether other industries should be similarly regulated will be considered in the next section.

2.2 Spare Parts Markets—Industries Other Than Automotive?

Because the spare parts market in the automotive industry is by far the most developed, if not the only competitive aftermarket, the debate surrounding design protection for must match spare parts is based almost exclusively on the technical

[3]Beldiman 2015, pp. 104 et seq.; Fischman Afori 2008, pp. 1105 et seq.

and economic realities of that industry.[4] In theory, however, demand for such parts could exist in other industries as well.[5] This prompts the question why in other industries this issue arises, if at all, only sporadically.[6]

This question was raised in the course of discussions preceding the adoption of the EU Design Directive, the automotive industry had suggested that other industries should be considered as well for purposes of regulating the spare parts issue. In response, the European Commission commissioned a report,[7] which in 2003 expressly concluded that only automotive spare parts are impacted by the repair clause: "Whilst the provisions of the Design Directive are generic in their application, they have elicited particular interest in the automotive sector, and cars especially, where the market for replacement parts is substantial. According to this study, no other sectors are specifically concerned".[8]

In contrast, a 2015 European study on the economic effects of design protection (EPEC Report) expressly includes other industries.[9] A detailed section on the repair clause[10] also discusses the watches and clocks industry, considered by the EPEC Report to be a highly design intensive industry.[11] From an economic standpoint, the Swiss watchmaking industry ranks highest, followed by Germany, Italy and France, all of which also feature relevant watch and clock manufacturers.[12] Control over replacement parts seems to be a concern also for this industry, albeit less acutely than in the automotive industry,[13] since the EPEC Report concludes: "Similarly to the automotive industry, the watch aftermarket consists of authorised and independent watch repairers. Since precise data on market concentration are not available, we are not able to discuss this issue in detail. However, what might be noted is that in a formal complaint to the European Commission, European confederation for watch repairer associations (CEAHR) claimed that authorised repairers act as monopolists in the respective aftermarkets for luxury watches as manufacturers refuse to supply parts to independents. Manufacturers, on the other hand, argued that their reasons relate to keeping high quality of services—as noted

---

[4]Cf. Riehle 2011, pp. 1 et seq.; see also European Policy Evaluation Consortium, Impact assessment of possible options to liberalise the aftermarket in spare parts – Final report to DG internal market (EPEC Report), Paris, November 18, 2003, pp. 1 et seq.; to this, see Straus 2005, pp. 973 et seq.

[5]Hartwig 2016, p. 108, who names coffee makers, vacuum cleaners, smartphones and tablets as other possible complex products whose spare parts would fall under the repair clause.

[6]E.g. printers, razor blades, watches. For watches and clocks, as well as and other complex products in general, see Europe Economics 2015, pp. 137, 142 et seq. See also Dyson Ltd v. Qualtex (UK) Ltd [2006] EWCA Civ 166 (on vacuum cleaners).

[7]EPEC Report 2003, p. i.

[8]EPEC Report 2003, p. i.

[9]See Europe Economics 2015.

[10]Europe Economics 2015, pp. 9, 146 et seq. (on visible spare parts).

[11]Cf. Europe Economics 2015, pp. 142 et seq.

[12]Europe Economics 2015, pp. 145.

[13]The graphic in Europe Economics 2015, p. 146, illustrates that a high share of spare parts for watches is produced by manufacturers from other countries, especially in Asia.

by the Commission, according to some of them 'up to 30% of repair work done in their after-sale services centres concern the damage caused by the inappropriate and wrongful repair executed by the watch repairers who do not possess proper knowledge and skills'.[14] While the issue of abuse of dominant position is still under investigation, the Report seems to indicate at least that authorised repair shops play an important role in that industry."[15]

However, beyond this conclusion, no reliable data are available and efforts to contact representatives of the watchmaking industry for further information, have remained unsuccessful.[16] This leaves the automotive industry as the sector with the most developed spare parts market, and accordingly the most significant interest in debates about legislation impacting it.[17] Several reasons might explain the prominence of the automotive aftermarket. In part, it is, of course, attributable to the considerable number of vehicles in circulation and the relatively high cost of acquisition of a vehicle compared to the cost of repair, which prompts repair over replacement. Even if measured by normal wear and tear, a fairly sizeable market would exist. With respect to visible spare parts, further factors stimulate the demand: first, the exposed position of must match parts in the context of their use (mobility) enhances normal wear and tear, and second, the frequency of non-probability events, such as traffic accidents. These factors account for the significant volume of the automotive spare parts market, as well as the fact that the data available derive almost exclusively from the automotive industry.[18] For this reason, the discussion in this book will center on spare part issues arising in that industry.

2.3 Spare Parts in the Automotive Industry

The automotive spare parts industry constitutes a significant market segment. In the EU alone, the spare parts business projected for 2020 is 230 billion Euro. Of this, visible parts hold a 25% share.[19] Regulation of spare parts will affect not only the

---

[14]Europe Economics 2015, p. 146, has taken this quote from General Court, 15 December 2010, Case T-427/08 – CEAHR v. Commission.

[15]Europe Economics 2015, p. 146.

[16]In preparation for the Roundtable discussion of 28 October 2015, representatives of the watch industry were contacted and invited to participate, but no affirmative response was received.

[17]The introductory part of the study made mention of other product categories, such as furniture, textiles, communication equipment. The absence of further investigation into this industry is to a large extent attributable to the unavailability of reliable data, given non-transparent market structures and the difficulty of differentiating between original and imitated spare parts. Cf. Europe Economics 2015, p. 17 et seq., p. 142.

[18]An added factor is the relatively high costs of automotive vehicles, which mandate repair by part replacement rather than a new acquisition.

[19]For more data, especially relating to the automotive market, see Reeves and Mendis 2015, p. 12; Berns 2013, pp. 14 et seq.

aftermarket itself, but will be felt throughout the vast supply chain underlying it.[20] These facts account for the intensive lobbying efforts by all stakeholders when it comes to spare parts regulation.

As a background for understanding the economic tensions arising in the automotive aftermarket, it is helpful to take a look at this market segment. Manufacture and distribution in the automotive spare parts business form a complex web of relationships. In order to achieve cost savings, timeliness of supply and ultimately, economies of scale, original equipment manufacturers (OEMs) outsource production, distribution and customer relations to spare part contract manufacturers. This has led to a division of the market of authorized spare parts into those which derive directly from the producer of the original vehicle[21] and those manufactured under contract by the so-called original equipment suppliers (OES). Spare parts made by OES are, while not manufactured by OEMs, authorized by them and presumptively subject to their quality control.[22] Elaborate contractual provisions are in place between OEMs and contract manufacturers, distributors and repair shops operating with the OEM's authorization. Spare parts which do not derive from OEM/OES sources are manufactured and distributed through networks of independent distributors and repair shops, also referred to as Independent Aftermarket (IAM).[23]

Structurally, the market can be divided into four levels: manufacturers, distributors, repair service and consumers. At the manufacturing level, spare parts may be produced by either OEMs, i.e. the developers and producers of the original vehicle, or by spare parts manufacturers, who may or may not be authorized by an OEM.[24] Non-authorized spare parts manufacturers, or independents are not contractually bound to the original manufacturer or its distribution network.[25] At the distribution level, distributors may be under contract to distribute spare parts from OEMs, from manufacturers authorized by OEMs, or alternatively from independent manufacturers. It is estimated that independent distributors make up about a 52%, and auto manufacturers a 48% market share. Spare parts manufacturers supply both distributors and auto manufacturers for purposes of distribution. According to surveys, the type of parts supplied by each segment is generally determined by vehicle age. OEMs supply primarily newer vehicles, 0–4 years old. The market for 4–8 year old vehicles is strongly disputed between the two, whereas the IAM tends to supply more parts for vehicles older than 8 years.[26] Finally, retailers may be under contract with original manufacturers or authorized distributors, or alternatively they might

---

[20]Cf. Legislative draft of the German government, BT-Drucks. 15/1075, pp. 65 et seq.; Drexl et al. 2005, pp. 449 et seq.; Pentheroudakis 2002, p. 685.

[21]Cf. Berns 2013, p. 11; for more details, see Blanken 2008, p. 24.

[22]See Riehle 1993, p. 52.

[23]Berns 2013, pp. 12, 16 et seq.; Blanken 2008, p. 25.

[24]For an overview and comparison of data from the different EU member states (from 2005), see Straus 2005, p. 978.

[25]Autorité de la concurrence 2012.

[26]Berns 2013, p. 19.

operate as IAM distributors and repair shops. The IAM segment is supplied by both authorized and independent suppliers, each holding an about 50% share of the market.[27] Finally, at the consumer level,[28] spare parts reach the final user by way of two channels: on the one side via the vehicle manufacturers, through the network of authorized dealers and repair shops (OEM/OES) and on the other, through the independent aftermarket. As a rule, production of parts by OEMs ceases after about 7 years, even though some auto manufacturers give 10-year warranties.[29] Longer term demand is driven by consumers who have an interest in having spare parts available for as many years as possible after purchase of the vehicle.[30] In the event original parts become unavailable through the OEM/OES channel, they are frequently procured from independent manufacturers.

Despite the complexity of these arrangements, the interests of the individual stakeholders are clearly delineated between OEMs and independent manufacturers. Their respective positions vis-à-vis regulation of this market segment can be outlined in general terms as follows.

2.4 Stakeholder Dynamics in the Spare Parts Business

2.4.1 Challenges to New Market Entrants

Independent spare parts manufacturers perceive themselves to be at a disadvantage in their ability to enter into and compete in the spare parts market. Because they normally lack access to the original design and associated know-how, they must reverse engineer parts. Often this involves extensive research and extraction of information on the precise structure of the original part and its properties, in order to ensure compatibility with the original product.[31] Production facilities must be established and distribution networks must be developed.[32] The increasing technical complexity of automobiles adds to the disadvantage of independent suppliers, as software diagnostic tools are costly to acquire and often proprietary to OEMs. Consequently substantial up-front investments are required in order to enter the

[27]Berns 2013, pp. 12, 16 et seq.; Blanken 2008, p. 25.

[28]In the EU, this fourth level consists of approx. 287 million users, i.e. automobile owners, with a market volume of about 765 billion Euro (in consumer prices), not including supplies such as oil, chemical and other minor components.

[29]Europe Economics 2015, p. 144; for details on the market structure (from the consumer's perspective), see also Drexl 2012, p. 377.

[30]For automobile manufacturers, it is complicated to permanently keep the spare parts ready (for details, see Kroher 1993, p. 464).

[31]Berns 2013, p. 16.

[32]Kur 2010, note 44.

market.[33] Independent suppliers are concerned that the OEMs' ability to secure IP protection on the design of spare parts would effectively exclude independents from the market.[34] Consequently, their position is to advocate liberalization of the spare parts markets, *inter alia* by eliminating design protection for the spare parts.

2.4.2 Challenges to the OEM Industry

OEMs have the advantage of already being "in" the spare parts market, having the benefit of production lines and distribution networks that were established in conjunction with the original product (the vehicle). However, it seems that the investment is rarely amortized through sales of those products. OEMs therefore must often rely on revenue from the spare parts market to recover their original investment, support future development and stay afloat in times of economic downturn. Competition from independent suppliers undermines this business model, largely because independent suppliers tend to concentrate spare parts for which production costs are low, and which are most profitable.[35] This not only reduces the OEMs' market share but burdens them with manufacturing and stocking, often for years, parts that are less or not profitable. All of this diminishes the resources available to OEMs for future development.[36] Of further concern are low quality counterfeits and the associated safety risks, which tend to have an adverse impact on OEM reputation. Consequently, OEMs advocate strong IP protection of spare parts.

2.4.3 Public Policy Concerns

Finally, from a public policy perspective, IP regulation of the spare parts market raises additional complex questions. On the one hand, eliminating IP protection in order to increase competition would lead to more favorable pricing of spare parts and benefit consumers. Such measures are supported by spare parts alliances and the insurance industry. On the other hand, national legislators, especially of

[33]Independent manufacturers have also raised concerns about the fact that the imbalance in economic strength between primary producers and contract manufacturers, the contractual provisions often contain provisions reflecting the tension in interests among these market participants, such as exclusivity obligations, minimum purchase, or single brand requirements. See Berns 2013, p. 19.

[34]Certain advances in the position of the independent suppliers have resulted from intervention of competition law limiting in the OEMs' ability to use restrictive contractual clauses in supply and distribution agreements, as well as based on withholding of technical information designed to eliminate or limit competition in the secondary markets. Riehle 1993, pp. 54 et seq.

[35]Berns 2013, p. 16.

[36]Cf. The Asian Patent Attorneys Association (APAA) 2009.

countries with strong automotive industries are under pressure to support national industries, given the considerable ramifications in terms of job creation, balance of trade and GDP. Public safety resulting from possibly sub-standard quality of independent spare parts is a further concern and measures addressing these issues are strongly supported by OEMs.

Overall, achieving a balanced solution that meets the needs of the triangulated interests of OEMs, spare parts manufacturers/insurance industry and consumers presents policymakers with a difficult challenge. Legislatures and courts in most countries are still in the process of seeking solutions.

The following chapters will focus on developments in this area, primarily in the EU and the U.S., but with occasional reference to approaches adopted by other jurisdictions, such as Australia, South Africa, Brazil and Japan.

References

Autorité de la concurrence (2012) Avis n° 12-A-21 du 8 octobre 2012 relatif au fonctionnement concurrentiel des secteurs de la réparation et de l'entretien de véhicules et de la fabrication et de la distribution de pièces de rechange (The French Competition Authority's report of 2012). http://www.autoritedelaconcurrence.fr/pdf/avis/12a21.pdf

Beldiman D (2015) Expressive dimensions in design: a question of incentive? In: Beldiman D (ed) Innovation, competition, collaboration. Edward Elgar Publishing, Cheltenham (UK), Northampton, MA (USA), pp 104–131

Berns E (2013) Marktmissbrauch auf Ersatzteilmärkten im deutschen, europäischen und US-amerikanischen Markt [Diss.]. EUL Verlag, Lohmar

Blanken C (2008) Wettbewerbsrechtliche und immaterialgüterrechtliche Probleme des Zubehör- und Ersatzteilgeschäfts. Nomos, Baden-Baden

Drexl J (2012) Counterfeiting and the spare parts issue. In: Geiger C (ed) Criminal enforcement and intellectual property: a handbook of contemporary research. Edward Elgar Publishing, Cheltenham (UK), Northampton, MA (USA), pp 369–385

Drexl J, Hilty R, Kur A (2005) Designschutz für Ersatzteile – Der Kommissionsvorschlag zur Einführung einer Reparaturklausel. GRUR Int 449–457

Europe Economics (2015) The Economic Review of Industrial Rights in Europe—Final Report, MARKT/2013/064//D2ST/OP, Jan 2015, commissioned by the European Commission). http://ec.europa.eu/growth/tools-databases/newsroom/cf/itemdetail.cfm?item_id=8844

European Policy Evaluation Consortium (2003) Impact assessment of possible options to liberalise the aftermarket in spare parts—Final report to DG internal market (EPEC Report), Paris, Nov 18, 2003. http://oami.europa.eu/en/office/pdf/spare.pdf

Fischman Afori O (2008) Reconceptualizing property rights. Cardozo Arts Entertainment L J 25:1105–1178

Hartwig H (2016) Spare parts under European design and trade mark law. GRUR Int 102–108

Kroher J (1993) EG-Geschmacksmusterschutz für Kraftfahrzeug-Ersatzteile. GRUR Int 457–465

Kur A (2010) Limiting IP protection for competition policy reasons—a case study based on the EU spare-parts-design discussion. In: Hilty R (ed) Research handbook on intellectual property and competition law, pp 313–344

Pentheroudakis C (2002) Die Umsetzung der Richtlinie 98/71/EG über den rechtlichen Schutz von Mustern und Modellen in den EU-Mitgliedstaaten. GRUR Int 668–682

Reeves P, Mendis D (2015) The current status and impact of 3D printing within the industrial sector: an analysis of six case studies. Research commissioned by the Intellectual Property

Office, UK. https://www.gov.uk/government/uploads/system/uploads/attachment_data/file/549044/Legal-Empirical-_Executive-Summary.pdf

Riehle G (1993) EG-Geschmacksmusterschutz und Kraftfahrzeug-Ersatzteile. GRUR Int 49–70

Riehle G (2011) Immaterialgüterschutz in Sekundärmärkten. Die „Ersatzteilfrage" – Präzedenz für einen Paradigmenwechsel? In: Bechtold S, Jickeli J, Rohe M (eds), Recht, Ordnung und Wettbewerb. Festschrift zum 70. Geburtstag von Wernhard Möschel. Nomos, Baden-Baden, pp 1075–1096

Straus J (2005) Ende des Geschmacksmusterschutzes für Ersatzteile in Europa? Vorgeschlagene Änderungen der EU-Richtlinie: Das Mandat der Kommission und seine zweifelhafte Ausführung. GRUR Int 965–979

The Asian Patent Attorneys Association (APAA), Design Committee (2009) Questionnaire, "Are Spare Parts Protected? Must Fit/Must match Provision in Design Law". http://www.apaaonline.org/pdf/APAA_56th_&_57th_council_meeting/designCommittee/1-APAA%20Designs%20Committee%20-%20Special%20Topics_Questionnaire%202009.pdf

Chapter 3
Design Protection Relating to Component Parts of Complex Products (Spare Parts) in the EU and the U.S.

3.1 Introduction

3.1.1 Structure of EU Design Laws

In the EU, the importance of design as an important sector of economic and cultural achievement began gaining increasing recognition in the mid-late 1990s. This led to proposals for legal protection[1] and materialized in the two pillar structure of European design law we know today[2]: at national level, laws enacted in accordance with the Directive 98/71/EC on the legal protection of designs (Design Directive),[3] and, at European level, the design law, Regulation No. 6/2002 on Community designs, also referred to as Community Design Regulation (CDR).[4]

Decades of debates on questions of principle preceded the enactment of this new design law structure.[5] It departs from the partly copyright-inspired and partly patent-inspired design laws of its member states and replaces the existing patchwork of discrepant national design laws with a new approach. The new concept, the

---

[1]Kur and Dreier 2013, p. 354.

[2]Riehle 1993, p. 50: "dual system". It should be noted that these pieces of legislation were developed by EU lawmakers with applicability to all member states at the time of enactment. In the event the UK ceases to be a member state ("Brexit") these provisions may or may not remain applicable to the UK.

[3]Directive 98/71/EC of the European Parliament and of the Council of 13 October 1998 on the legal protection of designs, L 289/28; see also Wandtke and Ohst 2005, p. 91. For introductions to the Directive, see Pentheroudakis 2002, pp. 668 et seq. (also on the implementation of the Design Directive in the member states); Kur 1998, pp. 977 et seq.

[4]Council Regulation (EC) No. 6/2002 of 12 December 2001 on Community designs, L 3/1.

[5]See Kelbel 1985, pp. 669 et seq.

D. Beldiman and C. Blanke-Roeser, *An International Perspective on Design Protection of Visible Spare Parts*, SpringerBriefs in Law,
DOI 10.1007/978-3-319-54060-3_3

"design approach," is premised, not on inventive/creative efforts as are patent and copyright, but on the recognition that design as an endeavor, serves an economic role.

The first pillar of this new system, the Design Directive, was tasked with harmonizing the member states' national laws in the spirit of this new design thinking. Its national implementations brought about significant changes in member states' laws.

German design law,[6] for instance, dates back to 1876. In its earliest version, it had been strongly influenced by copyright law,[7] which had been the only means for protecting industrial designs prior to the adoption of the design law.[8] Because industrial designs combine utilitarian and aesthetic features, designs often were unable to meet the requisite threshold of creativity for copyright purposes. The German design law was meant to remedy this shortcoming. It provided a low level of protection,[9] and, unlike copyright law, which rewards the creator's efforts, was intended to promote commercialization of products. Still, this law remained conceptually patterned upon copyright law.[10] Only after passage of the Design Directive in 1998,[11] did Germany move away from the copyright-based model, by adopting the "Law on legal protection of models and designs" (Geschmacksmustergesetz, GeschmMG).[12] A modernized version became effective January 1, 2014, the "Law on legal protection of design" (Designgesetz, DesignG).[13]

The approach of the Design Directive was then carried forward into the second pillar of European design law, the CDR.[14] The Regulation went into effect in 2002 and created a unitary and independent Europe-wide community design.[15] Originally

[6]Former Geschmacksmustergesetz (GeschmMG).

[7]For this reason it was originally named Copyright law on models and designs (original title: Gesetz betreffend das Urheberrecht an Marken und Modellen (Geschmacksmustergesetz)). For details on the legislative history, see Schicker and Haug 2014, pp. 726 et seq.; Kelbel 1985, pp. 669 et seq.

[8]Schicker and Haug 2014, p. 726; Kelbel 1985, pp. 669 et seq.

[9]See BGH, 22 June 1995, I ZR 119/93, GRUR 1995, 581 – Silberdistel.

[10]Cf. Schicker and Haug 2014, pp. 726 et seq.; Kelbel 1985, pp. 669 et seq.

[11]Directive Nr. 98/71/EC of the European Parliament and of the Council of 13 October 1998 on the legal protection of designs, L 289/28.

[12]Gesetz über den rechtlichen Schutz von Mustern und Modellen. For an introduction to this law, see Wandtke and Ohst 2005, pp. 91 et seq.

[13]For the main changes, see Rehmann 2013, pp. 215 et seq.

[14]Council Regulation (EC) No. 6/2002 of 12 December 2001 on Community designs, L 3/1.

[15]For details see Kur 2002, pp. 661 et seq. There are registered (Article 1(1)(b) CDR) and non-registered design rights (Article 1(1)(a)CDR); for an introduction see Rahlf and Gottschalk 2004, p. 827. Registration is performed by the EU Intellectual Property Office (EUIPO), formerly Office for Harmonization in the Internal Market (OHIM), Alicante, Spain (Article 2 CDR).

based on the same draft, the CDR and the Design Directive are consistent in content and handle the key aspects of design law in the same manner.[16]

3.1.2 U.S. Design Patent Law

Design protection has been available under U.S. law since 1842, to protect non-functional design by way of design patents. It emerged at the behest of the then Commissioner of the Patent Office, who sought to emulate similar laws in England and France.[17] Congress perceived that "giving certain new and original appearances to a manufactured article may enhance its salable value, may enlarge demand for it and may be meritorious service to the public"[18] and passed design law in order to give encouragement to the decorative arts.[19]

Upon adoption, design law was placed within the U.S. Patent Act[20] and under the oversight of the U.S. Patent and Trademark Office (USPTO),[21] the same agency that had authority over utility patents. This occurred less due to the conceptual proximity of the two types of protection, and rather due to the personal influence of the Commissioner of Patents, and explains the strong influence of utility patent law on design patents in terms of both general principles and procedures.[22] This fact also accounts for quite fundamental differences compared to EU design law, such as for instance, the fact that applications are examined under the same stringent process applied to utility patents. The somewhat burdensome process, with uncertain outcome makes the barrier to obtaining design protection in the U.S. quite high compared to its *sui generis* design protection system in the EU.

The following will discuss, seriatim, the aspects of design law that are most relevant to the protection of spare parts, including, the protection requirements and scope of protection, protection of component parts of a complex product, the visibility requirement and the exclusions applicable to spare parts, under EU and U.S. law, respectively.

---

[16]But cf. Kur 2002, p. 661: "Gleichlauf zwischen nationalem und Gemeinschaftsrecht"; Beier 1994, pp. 716 et seq.; Riehle 1993, p. 50.

[17]Hudson 1948, pp. 382 et seq.

[18]Gorham v. White, 81 U.S. 511, 524 (1871).

[19]Gorham v. White, 81 U.S. 511, 524 (1871).

[20]35 U.S.C. § 171.

[21]See http://www.uspto.gov.

[22]See also DuMont 2010, p. 531, for a history of the development of design patents and the non-obviousness requirement; see also, Mueller and Brean 2011.

3.2 Protection Requirements and Scope of Protection

3.2.1 EU Design Protection Requirements and Scope of Protection

EU design law protects the two or three-dimensional[23] appearance of a product—whether industrial or artisanal in nature—that results from characteristics of lines, contours, colors, shape, surface, structure of material of the product or its ornamentation.[24]

Protection is predicated on two requirements: (1) novelty and (2) individual character.[25]

The novelty requirement refers to the identity or quasi-identity of a design with designs already made public. A design shall be considered new if no identical design has been made available to the public.[26] The individual character requirement denotes a qualitative step,[27] i.e. the fact that the design must differ to some extent from the available wealth of forms; the relevant perspective for the assessment being that of an "informed user".[28] A design is therefore considered to have individual character if the overall impression it produces differs from the overall impression produced by any design which has previously been made available to the public.[29] Individual character is a function of the degree of freedom the designer enjoys in developing the design, i.e. the greater the design freedom, the broader the scope of protection, and the more substantial a difference between the protected design and an accused design is required, compared to cases of lesser design freedom.[30]

---

[23]The former German conceptual distinction between the two-dimensional ("Muster") and the three-dimensional ("Modell") design was abandoned in the 2004 version of the law (GeschmMG), the law which preceded the current Designgesetz (DesignG, see supra section 3.1.1); see Schicker and Haug 2014, p. 727. For a detailed treatment of further concepts of design law from an international perspective see Pentheroudakis 2002, p. 670.

[24]Article 3(a) Community Design Regulation and § 1 nr. 1 Designgesetz (DesignG).

[25]Article 4(1) CDR and § 2(1) DesignG.

[26]A non-exhaustive list of actions listed in § 5 DesignG under German law are deemed to have constituted disclosure in the sense of § 2(2) DesignG; for CDR, see Article 7 (Article 5(1)). A 12-month grace period is granted (Article 7(2) CDR, § 6 DesignG): Publication by the designer, its successor in title or an abuse in relation to the designer or his successor in title within a 12-month period preceding the date of filing of the application, will not destroy novelty. This allows the designer to test the product in the market without jeopardizing protectability. See Wandtke and Ohst 2005, p. 97; Kur 2002, p. 666.

[27]Unlike originality in copyright law, a design is not required to bear the stamp of the designer's personality. Nor does "individual character" denote a quality requirement either aesthetically in terms of the design's appeal to the eye, or in terms of the designer's skills (unlike copyright law, which requires above average designer's skill, etc.). See Kur 2016.

[28]Kur and Dreier 2013, p. 356.

[29]Article 6(1) CDR, § 2(3) DesignG.

[30]Cf. Wandtke and Ohst 2005, p. 97.

Both the CDR and the Design Directive provide for both registered and unregistered design rights.[31] Applications for registered designs are not subject to substantive examination.[32]

The right granted to the owner of a design, provides for protection against commercial exploitation of the same or a similar design by a third party. Infringement is determined based on the same standards as the requirements of protection: the perspective of an informed user of products of the type in question, with the degree of freedom of the designer having to be taken into consideration."[33] The scope of the protection conferred by a design right shall include any design which does not produce on the informed user a different overall impression.[34] This means that the protection has an exclusionary effect much like patent law, regardless of whether or not the design was copied or not, or whether the alleged infringer had knowledge of the existence of the design.[35]

In short, in the EU, a registered design holder would therefore be entitled to prohibit a third party from placing into the market a design that produces an overall impression that is not different from its own design.

3.2.2 U.S. Protection Requirements and Scope of Protection

The subject matter of a U.S. design patent may relate to the configuration or shape of an object, to the surface ornamentation of the object or both.[36] A U.S. design patent is granted for "new, original, and ornamental design" as well as non-obvious articles of manufacture.[37] These requirements must be understood as follows.

The threshold criterion for protectability is whether the design constitutes an "article of manufacture", in other words whether it is embodied into, or applied to, a

[31]Unregistered design rights will not be discussed further as of relatively marginal relevance for the present discussion.

[32]Registration under the CDR is available at the EUIPO with the following specified procedures. http://www.euipo.org. The right arises upon registration for a period of 5 years, which can be extended up to a maximum of 25 years. Challenges to validity can be raised at the EUIPO (by application to the Invalidity Division, Article 52 CDR). Validity of a registration will also be subject to challenge in actions for nullification in the courts of applicable jurisdiction. The effect of a declaration of invalidity of the design would be retroactive. See Article 19 CDR; Article 12 Design Directive (see, e.g. § 38 DesignG). Applications are not substantively examined as to comply with requirements for protection prior to registration (Recital 18 of CDR, cf. also Article 85(1) CDR).

[33]Kur and Dreier 2013, p. 358.

[34]Article 10(1) CDR; Article 9(1) Design Directive.

[35]See Kur and Dreier 2013, p. 358.

[36]USPTO 2015, § 1502.

[37]35 U.S.C. § 171.

man-made tangible object. The hallmark of a design patent is its ornamentality.[38] This element requires an appearance that is aesthetically pleasing, attractive, the product of an artistic conception, embellished or adorned, or distinguished by its grace of symmetry of form, in short, features that indicate that the design is not dictated by function alone.[39]

A design's ornamental features must further display the requisite novelty.[40] This is measured by the "average observer test": the overall appearance of the design in the eyes of an average, or ordinary, observer must be different from the appearance of any other single prior design.[41] The originality requirement is not to be understood in the same manner as originality for copyright purposes, but is primarily intended to avoid issuance of a patent on a design derived from a source other than the inventor.[42] Finally, to qualify for protection, a design must not be found obvious at the time the design was invented.[43] The non-obviousness inquiry focuses on whether the design would have been obvious to a designer of ordinary skill who designs articles of the type involved, specifically, whether one of ordinary skill would have combined teachings of the prior art to create the same overall visual appearance as the claimed design; rather than individual features.[44]

The well-established rule for determining the scope of protection of a design patent is "that if, in the eye of an ordinary observer, giving such attention as a purchaser usually gives, two designs are substantially the same, if the resemblance is such as to deceive such an observer, inducing him to purchase one supposing it to be the other, the first one patented is infringed by the other".[45] The resemblance must be evaluated in terms of the effect of prior art and of functionality on the designer's creative freedom and the space available in which to maneuver.[46] In a crowded field, with the patented design close to the prior art, creative room is limited. Differences, which at first sight appear to be minor, may assume greater importance.[47] If the accused design includes a particular feature present in the patented design that departs conspicuously from the prior art, the accused design is

---

[38]35 U.S.C. § 171. see also Beldiman and Beconcini 2015, pp. 551 et seq.

[39]OddzOn Products, Inc. v. Just Toys, Inc. 122 F.3d 1396 (Fed. Cir. 1997); Richardson v. Stanley Works, 543 F3.d 1288 (Fed. Cir. 2010).

[40]35 U.S.C. § 171.

[41]Gorham Co. v. White, 81 U.S. 511 (1871).

[42]Chisum 2015, § 23.05.

[43]35 U.S.C. § 271; Chisum 2015, § 23.05.

[44]See In re Borden, 39 USPQ2d 1524, 1527 (Fed. Cir. 1996); see also Beldiman and Beconcini 2015, p. 551; Mueller and Brean 2011.

[45]Gorham Co. v. White, 81 U.S. 511 (1871).

[46]Prior art will thus prevent non-expert observers from aggregating too many aspects of the design. See Tushnet 2012, pp. 419 et seq.

[47]Egyptian Goddess, Inc. v. Swisa, Inc., 543 F. 3d 665, 677 (Fed Cir 2008).

more likely found to be infringing.[48] Functional elements on the other hand, must be factored out prior to assessing the proximity in appearance between the protected and the accused design.[49]

3.3 Protection of Component Parts

A key to facilitating the protection of spare parts under both EU and U.S. law is the ability to protect a component part separate from the product into which it is embodied. Enabling such protection, essentially established spare parts as individual protectable entities. This provides rightholders with targeted means of securing protection and preventing imitation of individual components parts.

3.3.1 Protection of Component Parts in the EU

In Europe, national laws have historically diverged on the treatment of separate component parts of complex products. In Germany, for instance, individual protectability of spare parts was formally first recognized in the 1980s,[50] but as a practical matter the industry had sought and obtained protection of separate parts under the GeschmMG even earlier.[51] The laws of the UK, Denmark, the Benelux countries, Spain, and Portugal, on the other hand, did not accept the notion of independent part protection.[52]

When it came to negotiating a single position for purposes of the Design Directive, two camps emerged: one in favor of protection of separate component parts, which had successfully been used by automotive manufacturers in various countries in Europe, such as France and Germany, and the other against, arguing that such protection would serve the automotive industry to gain an undue advantage in the market.[53] The European Commission ultimately sided with the former, in order to ensure that the purpose of design law of protecting the appearance, the outside and visible form, of a product is fulfilled. The European Commission reasoned that "design protection cannot only be granted for "one-unit" products, such as a vase or a belt buckle, but also for "complex" products, whose

[48]Egyptian Goddess, Inc. v. Swisa, Inc., 543 F. 3d 665, 677 (Fed Cir 2008).

[49]OddzOn Products, Inc. v. Just Toys, Inc. 122 F.3d 1396 (Fed. Cir. 1997); Richardson v. Stanley Works, Inc., 597 F3.d 1288 (Fed. Cir. 2010).

[50]See BGH, 16 October 1986, I ZR 6/85, GRUR 1987, 518 – Kotflügel; Riehle, FS Möschel, 2011, p. 1081.

[51]Kroher 1993, pp. 460 et seq.: BMW has been protecting individual designs since 1951.

[52]BGH, 16 October 1986, I ZR 6/85, GRUR 1987, 518 – Kotflügel. France also accepted protection of parts; for details, see Berns 2013, p. 222; Beier 1994, p. 718.

[53]Drexl et al. 2005, pp. 449 et seq.; Beier 1994, pp. 716 et seq.

outer skin is composed of various components, such as the body of a car. As long as they are produced and marketed as new products, there is no difference between "one unit" and "complex" products. Both are subject to design law in the same way and the effects of design protection are identical".[54]

This approach is reflected in both the Design Directive and the CDR.[55] Both provide protection for designs incorporated in a product that constitutes a component part of a complex product.[56] A "complex product" in turn, is composed of multiple components which can be replaced permitting disassembly and re-assembly of the product.[57] It is such complex products that the repair clause for must match parts would apply to.[58]

3.3.2 *Protection of Component Parts in the U.S.*

US law defines the subject matter of a design patent as relating to the configuration or shape of an object, to the surface ornamentation of the object or both.[59] Because protectability of component parts is not expressly provided for in the statute,[60] historically, it was believed that the claim must be for the design embodied in or applied to an entire article of manufacture.[61] This changed in 1980 as a result of the *In re Zahn* decision,[62] where the court allowed a design patent applicant to claim the design for the shank portion of a drill bit, without having to claim the entire drill bit. The court reached this result, based on its interpretation of the wording of 35 U.S.C. § 171(a), which applies to designs "for" articles of manufacture, rather than "of" articles of manufacture.[63] This decision was precedential in allowing design patent protection for component parts of complex products.

---

[54]European Commission, Proposal for a Directive amending Directive 98/71/EC, 2004.

[55]Article 4(2) CDR; Article 3(3) Design Directive (see e.g. § 4 DesignG).

[56]"A design applied to or incorporated in a product which constitutes a component part of a complex product shall only be considered to be new and to have individual character: (a) if the component part, once it has been incorporated into the complex product, remains visible during normal use of the latter; and (b) to the extent that those visible features of the component part fulfil in themselves the requirements as to novelty and individual character" (Article 4(2) CDR; Article 3 (3) Design Directive). "Normal use" within the meaning of Article 4(2) CDR means use by the end user, excluding maintenance, servicing or repair work (e.g. Article 4(3) CDR).

[57]Article 3(c) CDR; Article 1(3) Design Directive (see e.g. § 1 nr. 3 DesignG).

[58]See Llewelyn and Baresi 2009, p. 9.

[59]USPTO 2015, § 1502.

[60]The first sentence of 35 U.S.C. § 171 reads "(w)hoever invents any new, original and ornamental design for an article of manufacture may obtain a patent therefor, subject to the conditions and requirements of this title" (emphasis added).

[61]Oake 2011.

[62]In re Zahn, 617 F.2d 261, 265 (CCPA 1980).

[63]35 U.S.C. § 171(a).

In practice protection of a component part is achieved by the technique known as "portion claiming", which allows a design patent to focus on only a part of a design by showing the features to be protected in solid lines and the remainder, for which protection is not claimed, in broken lines. Complete omission of non-claimed parts is also an option, because "the scope of the claimed design is limited to those surfaces 'as shown' in the application drawing(s) in the absence of any additional written disclosure".[64]

3.4 Visibility of Component Parts

Visibility is a further requirement that defines protectability of designs under both EU and U.S. design law. It limits protection to parts that are visible in the normal course of use of the product, and thus has a substantial narrowing effect on the type of protectable parts.

3.4.1 EU Law on Visibility of Component Parts

Both EU and U.S. law provide that component parts are protectable only if they are visible in the course of normal use of the design, based on the rationale that the effect of aesthetic features is aimed at the user in the normal course of use of the product. A design will not meet these requirements unless, once they have been incorporated into the complex product, remain visible during the latter's normal use.[65] "Normal use" means use by the end user, excluding maintenance, servicing or repair work.[66] Furthermore, its visible features meet the requirements of novelty and individual character.[67]

In order to establish protectability, i.e. to determine whether the visible part of the product is new and displays individual character, it is necessary to determine which parts remain visible during normal use and whether the visible features of that part produce on an informed user an overall impression different from that produced by another design which was made available to the public prior to the one

[64]USPTO 2015, § 1504.04.

[65]Article 4(2) CDR; Article 3(3)(a) Design Directive (see e.g. § 4 DesignG). Such parts pertain primarily to the chassis, see Wandtke and Ohst 2005, p. 97.

[66]Article 4(3) CDR; Article 3(4) Design Regulation (see e.g. § 1 nr. 4 DesignG); see also Berns 2013, p. 215.

[67]Article 4(2) CDR; Article 3(3)(a) Design Directive (see e.g. § 4 DesignG). Such parts pertain primarily to the vehicle body, see Wandtke and Ohst 2005, p. 97.

at issue.[68] In other words, only the part(s) of that product that are left visible during its normal use will determine the overall impression produced. This was illustrated by the General Court's holding in *Kwang Yang Motors v. OHIM (Honda)*.[69] In that case, a lawnmower engine was mounted so that it was only partially visible in the normal course of use. The court reasoned that a user, standing behind the lawnmower would see the engine from the top and therefore the upper side of the engine is primarily visible to him. Consequently, the upper side of the engine was the one that determines the overall impression produced by the engine.

3.4.2 U.S. Law on Visibility of Component Parts

The ornamentality requirement in U.S. design law implies that that a design must be visible in order to meet the protectability standards, even though this requirement is not expressly set forth.

The question was first addressed in the case of *In re Webb*,[70] involving a design patent covering an artificial hip replacement. The application was rejected because once implanted in the body, the replacement is no longer visible. This decision was reversed on the ground that the correct approach would be to ask whether the appearance of the article was "a matter of concern" to a purchaser, which in this case might have been health professionals and individuals involved in the procurement of the product. Doctors might select a hip implant based on its ornamental appearance, based on the display of the design in advertisements and at trade shows.[71] The proper rule to be applied, which is now recognized by the USPTO, is that normal use and therefore ornamentality must be based on visibility at some point between its manufacture and assembly, i.e. "a period in the article's life, beginning after completion of manufacture or assembly and ending with the ultimate destruction, loss, or disappearance of the article",[72] including all commercial uses of the article prior to its ultimate destination.

---

[68]T-11/08, Kwang Yang Motors v. OHIM (Honda), [2011], ECR II -0000.

[69]T-11/08, Kwang Yang Motors v. OHIM (Honda), [2011], ECR II -0000.

[70]In re *Webb*, 916 F.2d 1553 (Fed. Cir. 1990); see also International Seaway Trading Corp. v. Walgreens Corp. (Fed. Cir. Dec. 17, 2009).

[71]In re *Webb*, 916 F.2d 1553 (Fed. Cir. 1990); International Seaway Trading Corp. v. Walgreens Corp. (Fed. Cir. Dec. 17, 2009).

[72]In re *Webb*, 916 F.2d 1553 (Fed. Cir. 1990).

3.5 Exclusions from Protection

3.5.1 Exclusions from Protection Under EU Law

Of primary relevance in the context of spare parts are the exclusions based on functionality and interoperability.[73]

Article 8(2) CDR excludes from design protection "features of the appearance of a product which must necessarily be reproduced in their exact form and dimensions in order to permit the product in which the design is incorporated or to which it is applied to be mechanically connected to or placed in, around or against another product so that either product may perform its function".[74] These parts are also referred to as "must fit" parts, the rule as "must fit" exclusion. Examples of such automotive parts are, exhaust pipes or clutch sleeves,[75] whose appearance is dictated by technical considerations.[76] The purpose of this provision is to ensure interoperability and to avoid monopolization of spare parts that may harm innovation.[77]

"Must fit" parts are to be distinguished from "must match" parts[78] whose shape is dictated by a harmonious appearance of the overall product, rather than by functional considerations.[79] Must match parts include automotive spare parts,[80] such as visible parts of a vehicle's body, but also spare parts for household or

---

[73] § 3 DesignG lists certain exclusions, of which only § 3(1) nr. 2 (see Article 7 Design Directive) is relevant for present purposes; almost the same in Article 8(2) CDR. Moreover, Article 8(1) CDR provides for an exclusion of "features of appearance of a product which are solely dictated by its technical function". For further details relating to the state of the law prior to 2014, see Wandtke and Ohst 2005, pp. 97 et seq.

[74] § 3(1) nr. 2 DesignG respectively, Article 8(2) CDR.

[75] Brtka and Soetens 2015.

[76] As such they would fall under patent or utility model protection, cf. Pentheroudakis 2002, p. 671.

[77] Brtka and Soetens 2015; Steinberg, in: Büscher et al. 2015, DesignG, § 3 para 3; Wandtke and Ohst 2005, p. 97. Monopolization based on design rights would hinder development of products with comparable technical characteristics, cf. Pentheroudakis 2002, p. 671; see also the German government's legislative proposal, BT-Drucks. 15/1075, p. 34.

[78] § 3(1) nr. 2 DesignG respectively Article 8(2) CDR. § 3(2) DesignG points to § 3(1) nr. 2 DesignG ("Erscheinungsmerkmale im Sinne von Absatz 1 Nr. 2"), also Article 8(3) CDR to Article 8(2) CDR ("Notwithstanding para 2").

[79] Rehmann 2014, pp. 25 et seq.; Wandtke and Ohst 2005, p. 97; on the requirement of visibility see also Article 3(3)(a) Directive.

[80] Cf. Derclaye 2009, p. 31; Wandtke and Ohst 2005, p. 97.

electronic devices.[81] Non-visible parts, such as internal parts of a car engine do not fall into this category.[82]

Arguably, some parts fit into both the must fit and must match categories.[83] For instance, a fender is part of the vehicle's body and must therefore fit its precise dimensions (must fit). At the same time, however, it must stylistically match the overall appearance of the vehicle, say in the shape of the edges. While the edges are not a technical requirement, they are necessary in order to create a harmonious whole with the rest of the vehicle, and therefore constitute both a "must fit" part—needing to adapt to the chassis, and a must match part—needing to conform to the overall shape of the chassis.[84] Considerable debate has been generated relating to protectability of must match parts, and ultimately the Design Directive ended up not regulating their protection at all.[85]

3.5.2 Exclusions from Protection Under U.S. Law

US law accomplishes a comparable exclusion of technically dictated features by way of the functionality exclusion. Designs that are primarily dictated by function are excluded from protection. Functionality is approached in terms of the functionality—ornamentality dichotomy. The reasoning is that "when function dictates a design, protection would not promote the decorative arts, a purpose of the design patent statute".[86] In other words, if a design is dictated by its function, then it is not ornamental, and thus no appropriate subject matter for design patent protection.[87] An aspect of a design is considered functional if it is essential to the use or purpose of the article embodying the design or if the aspect affects the cost or quality of that article.[88] In determining whether a design is primarily functional or primarily ornamental, the claimed design is viewed in its entirety, for the ultimate question is

[81]Steinberg, in: Büscher et al. 2015, GGV, Article 4 para 16; Bulling et al. 2011, p. 46; on the elements and functions automotive body parts, see Eichmann 1997, p. 600. The difference between "must fit" and "must match" parts cannot be unequivocally based on the comparison between the English terms "fit" and "match" (cf. Stein 2013), but for practical purposes "must fit" can be read as "having to fit in terms of shape" whereas "must match" requires a fit from a visual perspective (cf. Pearsall and Hanks 2003). Riehle 1993, p. 62, includes both categories under the umbrella concept of parts that must conform in terms of shape.

[82]Beier 1994, p. 718; see also Lorenzen 2002, para E.2.b), relating to the requirement of visibility in Article 14 Directive.

[83]See also Riehle 1993, p. 68. It may be more appropriate to speak of the *property* of a part as being "must fit", respectively "must match".

[84]This example is taken from Riehle 1993, p. 62.

[85]Beier 1994, p. 718; see also Lorenzen 2002, para E.2.b).

[86]Avia *Group v. L.A. Gear*Calif., 853 F.2d 1557 (Fed. Cir. 1988).

[87]L.A. Gear v.Thom McAn 988 F.2d 1117 (Fed. Cir. 1993).

[88]OddzOn Products, Inc. v. Just Toys, Inc. 122 F.3d 1396 (Fed. Cir. 1997); Richardson v. Stanley Works, 543 F3.d 1288 (Fed. Cir. 2010).

not the functional or decorative aspect of each separate feature, but the overall appearance of the article." However, in case alternative designs exist i.e. there are several ways to achieve the desired function, the design is more likely to serve a primarily ornamental purpose, and to be protectable.[89]

The following summary of relevant aspects of design law in the EU and the U.S. indicates that despite differences in structure and approach of the respective design laws, spare parts of complex products are eligible for design protection in both jurisdictions. Broadly speaking, protectability is premised on the fact that component parts may receive individual protection, that they must be visible, non-functional and meet requirements of novelty and individual character. While interpretations by courts in different jurisdictions may vary somewhat, in essence, parts protectable in one of the jurisdictions will also be eligible for protection in the other, including for present purposes, automotive spare parts.

3.6 Tensions Resulting from Design Laws in the Spare Parts Context

The fact that, under modern design laws, manufacturers of complex products are able to protect a product's parts individually has enabled a significant advance in the spare parts market. This more nuanced design protection also underlies much of the current debate relating to must match parts. Still, the extent to which spare parts are protectable under design law remains subject to important limitations. The requirement of visibility in the course of the normal use of the product considerably limits the type of parts that are protectable. Considerations relating to interoperability require exclusion of functional parts from protection.[90] Parts such as oil filters, clutches, etc. are therefore excluded.[91] What remains protectable under design law are visible parts with aesthetic features, such as fenders, hoods, bumpers, quarter panels, doors, headlights.[92] Even so, the market volume of protectable spare parts is quite significant. It is estimated that in 2020 the market for must match spare parts will exceed 50 billion Euro in the EU alone.[93]

[89]OddzOn Products, Inc. v. Just Toys, Inc. 122 F.3d 1396 (Fed. Cir. 1997); Richardson v. Stanley Works, 543 F3.d 1288 (Fed. Cir. 2010).

[90]Cf. Kur 1993, p. 71; Riehle 1993, p. 61. Design protectability of spare parts was first recognized in BGH, 16 October 1986, I ZR 6/85, GRUR 1987, 518 – Kotflügel.

[91]Arguably, the effect of this provision is also to exclude consumables such as printer cartridges. See Eichmann 1997, p. 859.

[92]Whether these parts are indeed protectable, given their very limited design freedom, remains a highly controverted point, particularly in Europe. See Berns 2013, pp. 222 et seq.

[93]To illustrate the general order of magnitude, the projected spare parts market for Europe alone in 2020 is 230 billion Euro. Design protectable spare parts represent about a 25% share. Berns 2013, p. 19.

Faced with growing counterfeits of individual parts, OEMs increasingly resorted to design protection. As a result, design filings increased dramatically in the automotive industry.[94] This enhanced use of design protection, in turn, impacted the ability of independent part producers to compete in the market and prompted a push for market liberalization. The legislative provision specifically promoted in this context, also known as the "repair clause", would exempt design protectable spare parts from protection, to the extent they are used in the repair market.

The proposal of the repair clause prompted an intense debate among stakeholders. In support of the repair clause, independent parts producers argued that protection of individual spare parts creates a monopoly position for OEMs in the aftermarket. In opposition of the clause, the OEMs relied on their IP rights under design law, as necessary to combat counterfeiting. In this way, the repair clause, a relatively obscure provision of industrial design law, has become the center of a heated debate about control over a multi-billion Euro worldwide market.

The following sections will explain the stakeholders' respective legal positions asserted in support of and against adoption of the repair clause, the current status of the debate and extent of adoption in the EU and the U.S.

References

Andrew RP, Ferrill ED (2014) Using design patent protection for replacement parts. Corporate Counsel. http://www.finnegan.com/resources/articles/articlesdetail.aspx?news=0f2a81f3-7963-4fa8-a92a-3046e1ed30de

Beier FK (1994) Der Musterschutz von Ersatzteilen in den Vorschlägen für ein Europäisches Musterrecht. GRUR Int 716–732

Beldiman D, Beconcini P (2015) The final impression counts: seeking common ground in design patent infringement. 97 J Pat Trademark Off Soc 551

Berns E (2013) Marktmissbrauch auf Ersatzteilmärkten im deutschen, europäischen und US-amerikanischen Markt [Diss.]. EUL Verlag, Lohmar

Brtka R, Soetens, R (2015) Focus on the automotive industry: the protection of spare parts using Community designs. In: Lexology. http://www.twobirds.com/en/news/articles/2015/global/designwrites-mar/focus-on-the-automotive-industry-the-protection-of-spare-parts-using-community-designs

Bulling A, Langöhrig A, Hellwig T (2011) Geschmacksmuster – Designschutz in Deutschland und Europa mit USA, Japan, China und Korea, 3rd edn. Carl Heymanns Verlag, Köln

Büscher W, Dittmer S, Schiwy P (eds) (2015) Gewerblicher Rechtsschutz, Urheberrecht, Medienrecht, 3rd edn. Carl Heymanns Verlag, Berlin

Chisum DS (2015) Chisum on patents (17 volumes): a treatise on the law of patentability, validity and infringement, Albany, NY (USA)

[94]E.g. in the period 2009–2014 the USPTO issued over 1,700 design patents to the top five automakers alone. However, the number of applications has increased in other industries as well. In the 5-year period from 2009 to 2013, design patent applications increased by about 40%, significantly outpacing the 28% growth in utility parent applications during the same period (Andrew and Ferrill 2014).

Derclaye E (2009) Repair and recycle between IP rights, end user license agreements and encryption. In: Heath C, Sanders AK (eds) Spares, repairs, and intellectual property rights. Kluwer Law International, Alphen aan Den Rijn (Netherlands), pp 21–55
DuMont J (2010) A non-obvious design: reexamining the origins of the design patent standard. Gonzaga L Rev 45:531–609
Drexl J, Hilty R, Kur, A (2005) Designschutz für Ersatzteile – Der Kommissionsvorschlag zur Einführung einer Reparaturklausel. GRUR Int 449–457
Eichmann H (1997) Kein Geschmacksmusterschutz für must-match-Teile? GRUR Int 595–609
European Commission (2004) Proposal for a Directive of the European Parliament and of the Council amending Directive 98/71/EC on the legal protection of designs (COM(2004) 582 final – 2004/0203 (COD). http://www.eumonitor.eu/9353000/1/j9vvik7m1c3gyxp/vi7jgsymsdz3
Hudson TB (1948) A brief history of the development of design patent protection in the United States. J Pat Trademark Off Soc 30:380–400
Kelbel G (1985) Zur Verbesserung des Geschmacksmustergesetzes. GRUR 669–676
Kroher J (1993) EG-Geschmacksmusterschutz für Kraftfahrzeug-Ersatzteile. GRUR Int 457–465
Kur A (1998) Europäische Geschmacksmusterrechtsrichtlinie verabschiedet. GRUR Int 977–982
Kur A (2002) Die Auswirkungen des neuen Geschmacksmusterrechts auf die Praxis. GRUR 661–670
Kur A (2016) Ersatzteilfreiheit zwischen Marken- und Designrecht. GRUR 20–30
Kur A, Dreier T (2013) European intellectual property: texts, cases and materials. Edward Elgar Publishing, Cheltenham (UK), Northampton, MA (USA)
Llewelyn D, Barresi V (2009) Right holders' control over repair and conditioning. In: Heath C, Sanders AK (eds) Spares, repairs, and intellectual property rights. Kluwer Law International, Alphen aan Den Rijn (Netherlands), pp 1–19
Lorenzen B (2002) Designschutz im europäischen und internationalen Recht – Zur Anwendung und Auslegung internationalen und europäischen Designschutzrechts. LIT Verlag, Hamburg
Mueller JM, Brean DH (2011) Overcoming the 'impossible issue' of nonobviousness in design patents, Ky. L J 99:419–554
Oake RG (2011) Design patent perspective: understanding functionality in design patent law: part I. In: Intellectual property today. http://www.designpatentschool.com/assets/Oake_OCT11%20V2.pdf
Pearsall J, Hanks P (2003) The oxford dictionary of English, 2nd edn. Oxford University Press, Oxford (UK)
Pentheroudakis C (2002) Die Umsetzung der Richtlinie 98/71/EG über den rechtlichen Schutz von Mustern und Modellen in den EU-Mitgliedstaaten. GRUR Int 668–682
Rahlf S, Gottschalk E (2004) Neuland: Das nicht eingetragene Gemeinschaftsgeschmacksmuster. GRUR Int 821–827
Rehmann T (2013) Das Geschmacksmustergesetz wird modernisiert. GRUR-Prax 215–217
Rehmann T (2014) Designrecht, 2nd edn. München, C.H, Beck
Riehle G (1993) EG-Geschmacksmusterschutz und Kraftfahrzeug-Ersatzteile. GRUR Int 49–70
Riehle G (2011) Immaterialgüterschutz in Sekundärmärkten. Die „Ersatzteilfrage" – Präzedenz für einen Paradigmenwechsel? In: Bechtold S, Jickeli J, Rohe M (eds) Recht, Ordnung und Wettbewerb. Festschrift zum 70. Geburtstag von Wernhard Möschel. Nomos, Baden-Baden, pp 1075–1096
Schicker SC, Haug T (2014) Grundzüge des Designgesetzes. NJW 726–730
Stein G (2013) Gebrauchswörterbuch Englisch-Deutsch & Deutsch-Englisch. De Gruyter Mouton, Berlin
Tushnet R (2012) The eye alone is the judge: images and design patents. J Intell Prop L 19: 409–426
United States Patent and Trademark Office (USPTO) (2015) Manual of Patent Examination Procedures (MPEP), 9th edn
Wandtke AA, Ohst C (2005) Zur Reform des deutschen Geschmacksmustergesetzes. GRUR Int 91–102

Chapter 4
Legislative Efforts Relating to Design Law in the Context of Spare Parts

The debate over the "repair clause" has been going on for decades in different jurisdictions and continues to do so. The context of the debate is largely the same across jurisdictions and involves the same players and market conditions, and consequently the stakeholders' economic positions are substantially the same in all countries. Because the arguments are doctrinally most developed in the EU, the following draws largely on scholarship from the EU, primarily Germany. It provides an overview over the main arguments made in support of and against adoption of a repair clause.

4.1 Arguments in Support and Against Adoption of a Repair Clause

The debate on this issue stems from two conflicting ideologies on the role of intellectual property[1]: proponents of the clause rationalize an exception to design rights based on concepts of competitive need and consumer benefit,[2] opponents of the clause base their arguments on concepts of property rights and a strict view of the nature of IP rights.[3]

---

[1]Riehle, FS Möschel, 2011a, p. 1081.
[2]For detailed overviews see Berns 2013, pp. 229 et seq.; Ruhl 2010, Article 110, para 6 et seq.; arguing in support of the clause Riehle, FS Möschel, 2011a, p. 108; Drexl et al. 2005, pp. 449 et seq.
[3]Cf. Eichmann 1997, pp. 602, 608 et seq.; Beier 1994, p. 724; Kroher 1993, pp. 460 et seq.

D. Beldiman and C. Blanke-Roeser, *An International Perspective on Design Protection of Visible Spare Parts*, SpringerBriefs in Law,
DOI 10.1007/978-3-319-54060-3_4

4.1.1 Arguments in Support of a Repair Clause

The writings of numerous scholars have rationalized the need for adoption of a repair clause.[4] The arguments stem predominantly from the automotive industry.[5] They are based on the fact that, unlike in the market for automobiles (primary market), in the context of the market for spare parts (secondary market), the effect of IP rights, which are ordinarily considered as economically beneficial, operates differently and is detrimental to competition.

This results from the fact that protectability of component parts generates a peculiar market scenario: because the spare parts at issue must match the original vehicle, substitutability by alternative designs is not possible; e.g. a fender that differs in shape from the original one, is useless for the repair of the original vehicle and therefore unmarketable.[6] A competitor wishing to enter the market must make parts identical to the original, which, in turn would run afoul of IP protection and would entitle the IP owner to exclude the competitor from the market. Thus, for IP protected must match parts in a secondary market no competition is possible. Holders of the design right become the sole source of parts for "their" products and gain complete control over the market. Purchasers of the original complex product are "locked" into using parts produced by the original manufacturer.[7] The steadily increasing use of design protection continues to tighten the OEMs' control of the spare parts market.[8]

While IP rights alone do not confer market dominance,[9] under the circumstances described above, an IP right can give a rightholder full control over access to the market.[10] That is so because the relevant market coincides with the protected product, and therefore is the only one that can meet the requirement of marketability. This fact excludes any possible substitutability. Granting original producers design protection for spare parts of a complex product in a market without substitutability fully excludes competition and places such producers in a de facto monopoly position.[11] The effect of the IP right in this case would exceed the quantum of exclusivity ordinarily granted by IP rights.

---

[4]Drexl et al. 2005, pp. 449 et seq.; Pentheroudakis 2002, pp. 684 et seq. ("Spannungsverhältnis"); Eichmann 1997, pp. 601 et seq.; Kur 1996, pp. 879 et seq.; Kur 1993, p. 71, arguing that the economic policy considerations are of greater significance; Kroher 1993, p. 460 et seq.; contra Beier 1994, p. 724. In this regard antitrust law is also of importance (Articles 101 et seq. TFEU (ex-Article 81 et seq. EC)), see Eichmann 1997, pp. 602, 608 et seq.; Kur 1993, p. 75.

[5]Cf. Llewelyn and Baresi 2009, p. 9; Drexl et al. 2005, pp. 449 et seq.

[6]See Riehle, FS Möschel, 2011a, pp. 1075 et seq., also for further information (his fact is uncontroverted).

[7]Riehle, FS Möschel, 2011a, p. 1079.

[8]Efroni 2012.

[9]Berns 2013, p. 237.

[10]Berns 2013, p. 237; Riehle, FS Möschel, 2011a, pp. 1075 et seq.

[11]Drexl et al. 2005, pp. 449 et seq.; see also Riehle, FS Moeschel, 2011a, p. 1083.

Once the possibility of competition is ruled out, absent substitutability, the original product manufacturer is in a position to charge monopoly prices,[12] rather than merely receiving the design premium to which it is entitled to under IP laws.[13] This results in large measure from the fact that revenue from the secondary market translates into direct profit: the investment made into creating the particular design has already been made at an earlier time, in connection with the complex product and has been amortized. Once the amortization period has elapsed, from an economic standpoint, there is no justification for a continued monopoly in the form of IP protection.[14] In absence of competition, design protection leads to an overcompensation of the rightholder which facilitates cross-subsidization of endeavors unrelated to spare parts. Proponents of the repair clause promote the notion of "reasonable" compensation, i.e. compensation that a rightholder would receive in a fully competitive market.

The situation might be different if IP protection in the secondary market would provide an incentive to innovate.[15] However, in this market no innovation can be expected.[16] This is so because the parts at issue had been developed for use in the original product for the primary market (new car). Nor can the design be improved, as the parts must match the original product in terms of appearance and dimensions and because the designer is bound to pre-existing shapes.[17]

Ultimately in design protection, it can be argued that, whether the manufacturer is entitled to a "zone of exclusivity" as a reward for innovating, and to a design premium, is determined by consumer preferences.[18] These preferences are exercised at the level of the primary market, e.g. the preference for a particular automobile design.[19] They are reflected in the price charged by the manufacturer in the primary market. Because as to spare parts no preferences are exercised, no design premium should be due the manufacturer.

All of this leads to the conclusion that, from an economic standpoint, design protection for spare parts is undesirable. From a legal standpoint, the same conclusion can be reached by way of two doctrinal avenues. One argument is that no justification exists for IP protection of a design, once it reaches the spare parts market. As discussed above, this is due to the fact that the investment made into the

[12]European Commission, Proposal for a Directive amending Directive 98/71/EC, 2004.

[13]Drexl et al. 2005, p. 449 et seq.

[14]This is consistent with the overall consideration that in the spare parts market, the welfare effect of IP rights is outweighed by other factors that result from the constellation in which spare parts are marketed. See Berns 2013, p. 237.

[15]Drexl et al. 2005, pp. 449 et seq.

[16]Berns 2013, p. 237.

[17]Drexl et al. 2005, pp. 449 et seq.

[18]European Commission, Proposal for a Directive amending Directive 98/71/EC, 2004, para 4.1.4.

[19]This preference is exercised at the primary market (new cars) when the car is sold but not at the spare parts market, where the consumer has no choice of the spare part, it must match to the original. In the latter case, the vehicle manufacturer as a rightholder would receive a monopoly premium, instead of a design premium (primary market) (European Commission, Proposal for a Directive amending Directive 98/71/EC, 2004).

design of the parts has been amortized at the level of the primary market, a design premium has been collected and there is no room for innovation at the level of the secondary market. IP laws are not meant to provide investment protection. They merely give the owner an opportunity to earn a reward within a competitive market.[20] The second argument states that because no possibility of competition exists in the market, competition principles may be used to adjust the scope of IP protection. Whether a legislator should be influenced by considerations which pertain to competition law in regulating IP, is subject to debate.[21] In general, competition tools cannot serve to second-guess a legislator's decision in IP matters, as their use is of an exceptional nature and reserved to solving structural problems rather than individual issues.[22] However, it can be argued that where it is known in advance that certain circumstances tend to curtail competition, reliance on a competition rationale would be appropriate. Thus, if it is known that conferring design protection in the space of spare parts tends to give rise to a problematic situation from the competition standpoint, it would be appropriate to correct the problem at the outset, by addressing it at the level of IP protection.[23] This would avoid its developing into a more serious problem that must later on be corrected by way of competition tools.[24]

An anticipatory correction of a potentially anticompetitive situation might take the form of an outright exclusion of design protection for spare parts[25] or, alternatively, of a limitation to the scope of protection. The precise means is open to debate.[26] An exemption from protection directed specifically at the spare parts market for repairs, would constitute a more elegant and effective solution. It would be narrowly tailored to the secondary market and recognize the fact that the problem is not present in the primary market.[27] No consensus exists as to whether such correction would be consistent with national and international norms governing IP law.

[20]See Riehle, EWS 1996, pp. 361 et seq.

[21]Berns 2013, p. 237.

[22]Drexl et al. 2005, pp. 449 et seq.; see also Berns 2013, p. 237.

[23]Drexl et al. 2005, pp. 449 et seq.

[24]Drexl et al. 2005, pp. 449 et seq.

[25]This position has in the meantime been abandoned, cf. Riehle 1997, p. 362, compared to Riehle 1990, p. 929. The first question raised is whether design protection should extend to must match parts at all, given the very limited freedom of the designer in the context in which the design's shape is dictated by the complex product. Cf. e.g. Riehle 1993, pp. 62 et seq., arguing that intellectual property law already incorporates public policy considerations (59); contra Kroher 1993, p. 457; also Beier 1994, p. 721, according to whom a repair clause contradicts the nature of design law.

[26]Drexl et al. 2005, pp. 449 et seq.; Pentheroudakis 2002, pp. 684 et seq.; Eichmann 1997, pp. 601 et seq.; Kur 1996, pp. 879 et seq.; Kroher 1993, pp. 460 et seq.; Kur 1993, p. 75, arguing that economic policy considerations are of greater significance; contra Beier 1994, p. 724. In this regard antitrust law is also of importance, Article 101 et seq. TFEU (ex-Article 81 et seq. EC), see Eichmann 1997, pp. 602, 608 et seq.; Kur 1993, p. 75.

[27]Drexl et al. 2005, pp. 449 et seq.

4.1.2 Arguments Against Adoption of a Repair Clause

Opponents of the repair clause take a different view of both the IP and the competition analysis.

4.1.2.1 IP Law Arguments

IP laws are the result of careful balancing of economic and social interests, designed to have a long-term effect in rewarding investment into producing innovative design.[28] Existing IP laws as well as the underlying policy considerations opt against adoption of a repair clause for several reasons.[29]

First, a repair clause is inconsistent with IP laws' well-established purpose of incentivizing innovation. Its effect would be to penalize the creator or innovator by withdrawing IP protection, while at the same time rewarding competitors who have made no innovative contribution to the products at issue. Under such circumstances, a repair clause would, in effect, incentivize imitation rather than innovation.[30]

Second, curtailment of existing IP rights by adopting a repair clause would interfere with the economic policy considerations underlying IP laws. A curtailment of this nature might be permissible in limited circumstances, usually when countervailing higher-ranking public interests, such as health, or national security are at stake.[31] Economic interests and distribution of wealth considerations, however, do not justify second-guessing longstanding policy decisions relating to innovation. The need for lower spare parts prices in the secondary market does not fall into a cognizable category of circumstances sufficient to justify reducing an existing IP right. An undesirable precedent would be set. If existing IP rights can be diminished in the spare parts market, this method might prove appealing in other industries and circumstances as well, with far-reaching consequences for the IP system overall. The proper mechanism for regulating prices is not IP law, but competition law.[32]

---

[28]See generally, Landes and Posner 2003, pp. 11 et seq.

[29]For overviews, see Beier 1994, pp. 716 et seq.; Kroher 1993, pp. 457 et seq.

[30]Cf. Beier 1994, p. 729. The fact that much of the imitation occurs overseas and would impact employment, would also be raised as an objection to the repair clause.

[31]Cf. Article 8(1) TRIPS.

[32]See Oake 2011; Saidman 2009; Beier 1994, p. 729. Furthermore, the spare parts market is not a simple case of regulating prices by increasing competition. The insurance industry is a third stakeholder in this debate, and only competition law is equipped to address possible inefficiencies in this market. Virtually all collision damage is subject to insurance both in the EU and the U.S. While consumers is held to be the primarily interested party, any benefit to the consumer is merely secondary, possibly in the form of reduced premiums while the primary beneficiary of reduced spare part prices are the insurers.

Third, in assessing the need for greater or lesser IP protection, a broad, encompassing view of the economic dynamics impacted by the presence or absence of IP rights is required. All possible facets of the question must be taken into consideration. A narrow view, for instance, might perceive the spare parts market as one in which no innovation occurs. This, however, ignores the ongoing improvements which go into production methods, material quality, safety features and marketing for spare parts, such as for instance the introduction of a crumple zone in body parts,[33] an innovation financed in part through revenue from the spare parts market. Loss of a significant segment of the market and of the associated revenue might have a long-term adverse effect on investment in innovation in both the primary and the secondary market.

Similarly, a narrow view might consider that in the automotive industry, the investment in the original design is amortized in conjunction with the sale of the vehicle. The actual amortization for such products however fluctuates with the market. Factors such as economic slowdown, highly competitive markets and technological developments impacting the mobility industry may all impact the price chargeable in the primary market and, with it, the amortization period. For this reason, a manufacturer's development and profit calculation must extend beyond the primary market. Reducing the scope of protection absent conclusive data relating to the actual amortization period, might bring with it the unintended consequence of throttling the producers' ability to invest in further innovation.[34]

Fourth, some countries already have adopted non-statutory legal doctrines which except repairs from infringement liability.[35] Application of these doctrines is sufficient to address any problems arising in the secondary market. A balancing of interests, if necessary, can be adequately handled by the judiciary under the "repair doctrine". Legislative intervention would therefore be unnecessary.[36]

[33]Generally, it raises doubts whether design law is in fact the appropriate vehicle to introduce barriers to market entry. For a position critical of adopting safety and quality considerations as part of the discussion see Drexl et al. 2005, p. 456. An interesting example for safety concerns is the new active hood of Mercedes-Benz, which in a collision with a pedestrian is capable of being raised, thus increasing the crash space. This considerably diminishes the risk of injury to the pedestrian (see http://m.mercedes-benz.de/techcenter/active_hood/detail.html). In contrast, hoods produced by independent manufacturers may lack this feature, making them a greater risk for pedestrians (see http://m.mercedes-benz.de/techcenter/active_hood/detail.htm).

[34]See also Beier 1994, p. 730.

[35]Aro Mfg. Co. v. Convertible Top Replacement Co., 365 U.S. 336, 342, 81 S. Ct. 599, 602, 5 L. Ed. 2d 592, 128, 128 U.S.P.Q. 354 (1961) (Aro I).

[36]Aro Mfg. Co. v. Convertible Top Replacement Co., 365 U.S. 336, 342, 81 S. Ct. 599, 602, 5 L. Ed. 2d 592, 128, 128 U.S.P.Q. 354 (1961) (Aro I); see also American Intellectual Property Law Association (AIPLA), Draft Resolution on the PARTS bill (AIPLA Resolution), 2015.

Finally, a repair clause is said to violate international law norms.[37] Article 26(2) TRIPS prohibits exceptions or limitations to the scope of industrial design right that would adversely impact the legitimate interests of the rightholder or interfere with the normal exploitation of the rights.[38] The repair clause falls short of these requirements.[39] First, the rightholders' legitimate interests would be impacted, because owners of design rights would be deprived of the substance and use of their exclusive rights in the secondary market.[40] Second, the normal exploitation of an IP right entails the right to exclude all forms of competition that would detract significantly from the economic return anticipated from a grant of exclusivity.[41] A repair clause would impair this right, as it would prevent the rightholder from excluding competition in this particular market.[42] Further, the requirement of a 10 year protection under Article 26(3) TRIPS, is also violated, as the repair clause eliminates protection for a given market segment altogether.[43] In short, a repair clause is irreconcilable with the mandates of international law under the TRIPS agreement.

4.1.2.2 Competition Law Arguments

Opponents of the repair clause assert that, in general, manipulating the structure of intellectual property rights by way of exemptions, such as the repair clause, may have unintended consequences on innovation and is unlikely to appropriately address market-specific competition issues.[44] Neither IP nor competition law can justify adoption of the repair clause.[45] For competition law to intervene, abuse of a

[37]For details, see Beier 1994, pp. 728 et seq.

[38]Beier 1994, p. 728. In copyright law, Article 13 TRIPS contains the so-called *three-step test*, which is at least comparable to Article 26(2) TRIPS. But, while Article 13 TRIPS was implemented to secondary law (Article 5(5) of Directive 2001/29/EC of the European Parliament and of the Council of 22 May 2001 on the harmonisation of certain aspects of copyright and related rights in the information society (*InfoSoc Directive*; for details, see Senftleben 2010, pp. 76 et seq.), Article 26(2) TRIPS was not. However, despite that, both EU and the national legislators are bound to Article 26(2) TRIPS. See Cornels 2015, pp. 76 et seq., also to the rather theoretical question of whether design law actually contains a three-step test.

[39]Straus 2005, p. 971.

[40]Beier 1994, p. 728.

[41]Canada Pharmacutical Patents, WTO Dispute Resolution Proceeding, DS 114 (2000). While this proceeding refers to patent law and interprets Article 30 TRIPS, both the wording and the intent of Article 30 and 26(2) TRIPS can be considered as similar. See Straus 2005, p. 970.

[42]Straus 2005, p. 971.

[43]Beier 1994, p. 728.

[44]Europe Economics 2015, p. 155.

[45]Beier 1994, pp. 731 et seq.; Kroher 1993, p. 465.

dominant position is required.[46] The CJEU's competition jurisprudence makes it clear that the mere exercise of exclusive rights by a rightholder does not constitute an abuse of a dominant position.[47]

In determining whether a dominant position exists in a secondary market for spare parts, both the primary market, for the complex product, and the secondary market, for replacement parts, are relevant.[48] The determining factor is whether consumers have the ability to make an informed product choice in the primary market, i.e. for instance, whether they have available information to assess the cost of replacement parts over the life-cycle of a particular complex product compared to another product.[49] In the event they do, European Commission competition decisions have held that the proper market to consider for purposes of dominance is the product's overall lifecycle, i.e. the primary and secondary markets combined. This rule applies as long as the primary market is fully competitive,[50] which is the case

[46]CJEU, 29 April 2004, Case C-418/01 – IMS Health GmbH & Co. OHG v. NDC Health GmbH & Co. KG, [2004] I-5039; CJEU, 6 April 1995, Joined cases C-241/91 P and C-242/91 P – Radió Telefis Éireann (RTE) and Independent Television Publications Ltd (ITP) v. Commission of the European Communities (Magill) [1995] I-743. In general, EU competition provisions are more stringent than those applicable under U.S. law (e.g. Sherman Antitrust Act). OEMs take the position that the spare parts market is dominated by a coalition between insurance companies and spare parts manufacturers, such as Progressive and LKQ, which at least in the U.S. control the market of collision repairs and affirmatively promote service providers who sell spare parts, "steering" consumers to providers in which they own shares.

[47]CJEU, 5 October 1988, Case C-238/87 – AB Volvo v. Erik Veng (UK) Ltd. [1988] 6211. For instance the court also held that refusal to license is an exercise of an exclusive right. This holding may be questionable in light of later jurisprudence. Cf. also Kroher 1993, p. 463.

[48]More recent studies have rejected the method of considering markets separately for purposes of competition analysis: "The notion of market separation must be rejected as an unconvincing competition analysis that is easily contradicted by standard findings in EU competition cases" (Europe Economics 2015, pp. 154 et seq.).

[49]In cases such as European Commission, Case Nr. IV/34.330 – Pelikan v. Kyocera and European Commission, Case Nr. IV/E 2/36.431 – Info-Lab v. Ricoh, the Commission developed criteria to determine if a manufacturer holds a dominant position in the secondary market for compatible products. The criteria used by the Commission in the Info-Lab v. Ricoh case were whether (1) the customer can make informed choice including lifecycle pricing, (2) the consumer is likely to make such an informed choice, (3) in case of an apparent policy of exploitation being pursued in one specific aftermarket, a sufficient number of customers would adapt their purchasing behavior at the level of the primary market and (4) the above all applies within a reasonable time. See also Europe Economics 2015, p. 155.

[50]European Commission, Case Nr. IV/34.330 – Pelikan v. Kyocera. The Commission decided that Kyocera, the producer of printers and toners, did not hold a dominant position in the market for toners, because it did not hold a dominant position in the primary market for printers.

in the automotive industry. Therefore, as relates to automotive spare parts, the relevant market must be the product-plus-lifetime-repairs market.[51] An analysis for purposes of competition law cannot rest on the secondary market alone, and must instead include the primary market.

Specifically with respect to the auto spare parts market, the CJEU, in the CICRA *v. Renault*[52] and *AB Volvo v. Erik Veng*[53] decisions listed three types of conduct which would violate competition law: arbitrary refusal to supply independent repair service providers, inappropriately inflated prices and refusal to produce parts for a certain model.[54] The CJEU refused to find that charging higher prices for protected parts constitutes an abuse for purposes of competition law, recognizing that manufacturers must amortize their development costs.[55]

4.1.2.3 Other Arguments

A number of other arguments that are not directly related to IP law have been asserted in opposition to a repair clause. These fall into the general categories of unjustified deprivation of constitutionally guaranteed property rights,[56] unequal treatment by way of a sectoral regulation that favors certain industries at the

[51]Europe Economics 2015, p. 155.

[52]CJEU, 21 June 1988, Case C-53/87 – Consorzio Italiano della Componentistica di Ricambio per Autoveilici (CICRA) and Maxicar v. Régie Nationale des Usines Renault [1988] ECR 6039, 237.

[53]CJEU, 5 October 1988, Case C-238/87 – AB Volvo v. Erik Veng (UK) Ltd. [1988] 6211.

[54]CJEU, 11 May 2000, Case C-38/98 – Régie nationale des usines Renault SA v. Maxicar SpA and Orazio Formento [2000] I-2973; CJEU, 5 October 1988, Case C-238/87 – AB Volvo v. Erik Veng (UK) Ltd. [1988] 6211.

[55]CJEU, 11 May 2000, Case C-38/98 – Régie nationale des usines Renault SA v. Maxicar SpA and Orazio Formento [2000] I-2973; CJEU, 5 October 1988, Case C-238/87 – AB Volvo v. Erik Veng (UK) Ltd. [1988] 6211; LG Hamburg, 18 September 2015, 308 O 143/14, GRUR-RS 2015, 16872; to this, see Klawitter 2015, p. 466.

[56]IP is explicitly protected under Article 17(2) of the Charter of Fundamental Rights of the European Union. It also falls under the term "property" e.g. in Article 14 of the German constitution (Grundgesetz). Unjustified taking of property is unconstitutional. In the U.S. it is argued that enactment of a repair clause would represent an unconstitutional taking under the 5th and 14th Amendments, in depriving design patent owners of the benefits of their property rights in their design patents (see AIPLA Resolution 2015).

expense of others, as the repair clause adversely impacts primarily the automotive industry,[57] improper application of the aesthetic functionality doctrine,[58] as well as safety concerns.[59]

4.1.3 *Summary of the Arguments*

In summary, the legal arguments surrounding the adoption of a repair clause boil down to a doctrinal debate of several issues, including the role and function of IP law, the primacy of competition law over IP law and the respective ability of these laws to control whether the repair clause is a justifiable exemption or not. Because strong arguments support both positions, no easy answers exist. Ultimately, the outcome cannot hinge on legal doctrine but is likely determined based on economic policy, specifically on whether in the perception of a national legislator the welfare effect of IP rights is outweighed by other factors that result from the constellation in which spare parts are marketed.[60] The debate will play out in both legislatures and courts and solutions will vary among jurisdictions. The following will discuss how these debates have unfolded in the EU and the U.S., as well as their outcomes.

4.2 The Repair Clause in the European Union

As described earlier, European design law is based on a two-pillar system: the national pillar governed by the Design Directive and the EU level governed by the CDR. At national level, the Design Directive is meant to achieve harmonization

---

[57]Also, the equal protection argument holds that an exceptional regulation limiting IP rights in the public interest would have to be done across industries and not by way of regulation favoring the spare parts and insurance industries, cf. Llewelyn and Baresi 2009, p. 9; Drexl et al. 2005, pp. 449 et seq.

[58]That the aesthetic functionality doctrine denies design protection altogether is wrong, because aesthetic features for which no design alternatives exist are the very essence of design protection. It is therefore wrong to assert that protection would be barred by the doctrine of aesthetic functionality. See Beier 1994, p. 721: "Design protection must not be limited by aesthetically functional considerations (note: as opposed to technically functional ones, for which there are no alternatives). A shape that is dictated by aesthetic reasons is the very concept of a protectable design. This is also true for designs which constitute component parts of a whole, whose aesthetic function allows no creative freedom for alternative designs, without changing the appearance of the whole. A must match exception, i.e. the exclusion of design protection for parts of a product which contribute to its overall appearance together with other products and are determined only by aesthetic considerations is contrary to the nature of design protection. It excludes the very core of the protectable design".

[59]Cf. e.g. Beier 1994, p. 730; Kroher 1993, p. 464. Many question whether design law is in fact the appropriate vehicle to introduce barriers to market entry and address safety concerns. For details, see supra section 4.1.2.1.

[60]Berns 2013, p. 237.

among member states. However, with respect to design protection of must match spare parts, the Design Directive fails to do so. As a result, adoption of the repair clause at national level among EU member states diverges. For instance, Belgium, Hungary, Ireland, Italy, Latvia, Luxemburg, the Netherlands, Poland, Spain and the UK have adopted provisions that substantially parallel the repair clause of Article 110(1) CDR in their national legislation, while countries such as Austria, Denmark, Finland, France, Germany, Portugal and Sweden have not.[61] At EU level, on the other hand, Article 110(1) CDR provides for a repair clause that exempts must match parts from infringement liability and is assertable as a defense to alleged infringements of designs protected under the CDR. This split in approach between national laws and the CDR has been accepted[62] in the hope that future European Commission proposals will result in changes that eliminate the difference.[63] The events leading up to these diverging provisions will be discussed next.

4.2.1 *The European Design Directive*

4.2.1.1 History and Debate Surrounding the Repair Clause in the Design Directive

When in the late 1990s the EU legislature sought to harmonize design laws within Europe, the issue of spare parts protection proved one of the most contentious. The European Commission seemed inclined to liberalize the spare parts market in order to allow third parties access to the market. Its proposed method was to exempt from design protection must match spare parts, in the event they are used for purposes of repair to restore the product to its original appearance.

To this end, the European Commission presented successive versions of the repair clause. The first version appeared as Article 14 of the Directive Proposal 1993, in the same form as Article 23 of the Regulation Proposal, and provided that must match parts, if used for repair purposes, must be free from protection for three years after the date of their placement on the market.[64] No agreement was reached on this version. A 1996 proposal eliminated the three year duration and, instead, made the exemption contingent on certain disclosures, such as the product's origin,

---

[61]Graphic representations of same are to be found in Wahlen and Kloster 2012, p. 6; Riehle, ECAR, 2011b, p. 8.

[62]Background in Kur 1998, pp. 981 et seq. Also cf. Recital 13 of the CDR; to this, Wandtke and Ohst 2005, pp. 97 et seq.

[63]In any event, a change of Article 110(1) CDR is not expected in the near future. See Ebert-Weidenfeller 2014, p. 56.

[64]European Commission, Proposal for a European Parliament and Council Directive on the legal protection of designs of 3 December 1993, COM (93) 344, ABl. 315/14, available at: http://aei.pitt.edu/9095; see also Berns 2013, note 1969; for details, see Ohlgart, in: Franzosi 1996, pp. 146 et seq.

informing rightholders about the intended use of the part, offering reasonable remuneration,[65] as well as providing regular and reliable information relating to the volume of use.[66] Again the European Commission and the Parliament were unable to reach an agreement. Because the spare parts issue threatened to delay adoption of the Design Directive, an interim compromise solution was adopted, in the form of Article 14 of the Design Directive.[67]

Article 14 of the Design Directive provides that national states are to maintain in force their existing national regulations until a Commission proposal made in accordance with Article 18 Design Directive is adopted. Article 18 requires that the Commission submit, within three years, "an analysis of the consequences of the provisions of this Directive for Community industry, including the industrial sectors which are most affected, particularly manufacturers of complex products and component parts, for consumers, for competition and for the functioning of the internal market." The article further required the Commission to make substantive proposals one year thereafter.[68] Pending this, member states were entitled to maintain their existing applicable laws, with the proviso that if a country wishes to make changes to its existing laws, such changes would have to be in the spirit of liberalization of the spare parts market. The compromise solution of Article 14 Design Directive therefore merely operated to maintain the *status quo* and to postpone action to a later date, which is why it is also referred to as "freeze-plus" provision.[69]

Following adoption of the Design Directive, the Commission acted on the spare parts question by ordering an examination of the issue and preparation of a report, which was published in November 2003 and titled "Impact assessment of possible options to liberalise the aftermarket in spare parts" (EPEC Report).[70] The conclusions listed in the EPEC Report and the Commission's own evaluation of the impact of the Design Directive prompted the conclusion that a repair clause should be adopted.[71] As a result, on 14 September 2004, the Commission submitted yet

[65]The draft version left the specific measures to be determined by the states: "Member states shall ensure that consumers are duly informed about the origin of spare parts so that they can make an informed choice between competing spare parts". This would have allowed states to use more flexible means, such as unfair competition law to enforce this requirement.

[66]European Commission, Amended Proposal for a European Parliament and Council Directive on the legal protection of designs of 21 February 1996, COM(96) 66, available at: http://aei.pitt.edu/10041; see also Berns 2013, note 1969.

[67]For more details, see Kur 2016, p. 21.

[68]Wandtke and Ohst 2005, p. 97; for a detailed discussion see Kur 1998, p. 977.

[69]Cf. Llewelyn and Baresi 2009, p. 8; Wandtke and Ohst 2005, p. 98; Kur 2002, p. 669. This regulation is also referred to as "stand still plus", see also Pierson, in: Pierson et al. 2010, § 38. IV.4. Also discussed was a "free for all" principle, allowing members to choose their own solutions. See Berns 2013, p. 224; Kur 1999, p. 626. Kur 1998, p. 977, expresses concerns with respect to the exhaustion of rights in design law under Article 15 Design Directive.

[70]EPEC Report, 2003.

[71]See European Commission, Proposal for a Directive amending Directive 98/71/EC, 2004, pp. 25 et seq.; cf. Straus 2005, pp. 965 et seq., 973, also with details regarding the EPEC Report.

another proposal for a repair clause.[72] in wording very similar to that of Article 110 (1) CDR,[73] i.e. excluding design protection for must match parts if certain prerequisites are fulfilled.[74]

This proposal has never been accepted.[75] Accordingly, the Design Directive has not been amended to add a repair clause and continues to operate under the freeze-plus regime of Article 14. Member states are free to maintain their own laws governing must match parts.[76] In the meantime, the Commission has officially withdrawn its proposal on 21 May 21 2014, following the veto of several countries in the European Council in 2008.[77] Further reports were commissioned, the first of which was issued in 2015.[78]At national level, the debates about adoption of a repair clause are ongoing in several countries, such as Germany,[79] France,[80] Romania,[81] etc. The situation in the most prominent "adopters" and "non-adopters" of the clause is briefly described below.

4.2.1.2 The Repair Clause in European National Laws

4.2.1.2.1 Italy

One of the more prominent adopters of the repair clause under the Design Directive is Italy. Given the economic importance of Italy's independent spare parts industry, the Italian legislature considered liberalization of the spare parts early on.[82] In 1994, the Italian competition authority also demanded liberalization of the spare parts

[72]For details, see European Commission, Proposal for a Directive amending Directive 98/71/EC, 2004.

[73]See infra section 4.2.2.

[74]See European Commission, Proposal for a Directive amending Directive 98/71/EC, 2004.

[75]At least, the European Parliament had generally approved it in 2007. See also Kur 2016, p. 21.

[76]See also LG Düsseldorf, 28 November 2013, 14 c O 304/12 U, GRUR-RS 2014, 00345. In 2007, Germany already had a concrete legislative proposal based thereon, cf. Choudhry 2014.

[77]These countries were: Czech Republic, France, Germany, Romania, Slovakia and Sweden (Hartwig 2015; Schriever et al. 2014; Wahlen and Kloster 2012, p. 2). See also Hartwig 2016, p. 102; Kur 2016, p. 21.

[78]Europe Economics 2015.

[79]Schriever et al. 2014; Wahlen and Kloster 2012, p. 2.

[80]See generally, Autorité de la concurrence 2012.

[81]See generally, Autorité de la concurrence 2012.

[82]According to Guizzardi 2005, p. 302, the clause is an expression of a competition-friendly approach of the Italian legislator, who wishes to limit the exclusive right based on the need for protecting competition. A further reason may be that the Italian legislator is generally quite Europe-friendly and sought to act consistently with Europe's goal to liberalize the market for spare parts. Thus, Italy normally is an early implementer of European directives, often by adopting their literal wording.

aftermarket.[83] Courts in the meantime dealt with the issue under unfair competition law and subsequently under design and trademark law. Two decisions by the Italian Corte di Cassazione, from 1996 and 2001, respectively, placed the protectability of spare parts in question.[84] These developments made Italy a receptive environment for the Directive.

The repair clause was finally implemented in 2001 in Article 241 Codice della proprietà industriale (CPI),[85] as a transitional provision, effective until a change of the Design Directive pursuant to Article 18. The tenor of Article 241 CPI is generally consistent with Article 110(1) CDR.[86] Unlike Article 110 (1) CDR however, the Italian repair clause is phrased more broadly, providing a defense against IP laws in general, rather than merely against design law.[87] The scope to be given to Article 241 CPI has been debated. Some have argued for an interpretation encompassing and possibly exceeding the scope of Article 110(1) CDR. The better view is however to construe Article 241 CPI in conformity with European law, i.e. in light of Article 14 Design Directive.[88] The fact that the Directive, unlike the CDR, contains no repair clause,[89] speaks initially against interpreting national law in the same way as the CDR. Still, a similar interpretation can be reached by way of comparing the respective recitals of the Directive and the CDR. Both Article 14 and Recital 19 of the Design Directive utilize the term "design". In this they are consistent not only with Article 110(1) CDR, but also with Recital 13 of the CDR, which makes express reference to the Design Directive, stating that "Full-scale approximation of the laws of the Member States on the use of protected designs for the purpose of permitting the repair of a complex product (…) could not be achieved through Directive 98/71/EC". Unlike other regulations such as Article 16 CDR, this recital does not use the term "exclusive rights", which would refer to all

[83]Autorità Garante della Concorrenza e del Mercato, AS29 – Normativa brevetti per modelli e disegni ornamentali per le parti staccate della carrozzeria delle automobili, 3 August 1994, in: Bollettino 2004, Nr. 32–33; see also Guizzardi 2005, pp. 301 et seq.

[84]Cass. 26 July 1996, Nr. 6644 – Hella v. Aric. For a German translation, see GRUR Int 1997, p. 650 et seq., with comments by Zorzi; for an analysis, see Eichmann 1997, p. 595 ff. See also Cass. 3 January 2001, Nr. 61. For a German translation, see GRUR Int 2002, 943, with comments by Fabbio.

[85]This was effected by law Nr. 95/2001, cf. Guizzardi 2005, p. 299. For details on the case Fittante 2001, pp. 532 et seq.

[86]Article 241 CPI: "Until the Directive 98/71/EC of the European Parliament and of the Council of 13 October 1998 on the legal protection of designs will be modified at proposal of the Commission according to its Article 18, the exclusive rights on the components of a complex product cannot be asserted in order to hamper the manufacture and the sale of components that look the same for the repair of the complex product, in order to restore its original aspect". See also Hackbarth 2015, p. 481; Guizzardi 2005, p. 300.

[87]Article 241 CPI creates a right to reproduce components of a complex product which are destined to restore the complex product to its original condition, despite the existence of intellectual property rights.

[88]Cf. also Würtenberger and Loschelder 2015, p. 349.

[89]See supra section 4.2.1.

IP rights.[90] Similarly, Article 96(1) CDR emphasizes that the provisions governing other IP rights remain unaffected, and thus shape the sphere of application of design laws accordingly.[91] Consequently, repair clauses of national laws must be construed in accordance with EU law and with Article 110(1) CDR.[92] This approach is consistent with the tenor of the CJEU opinion in *Ford v. Wheeltrims.*[93]

4.2.1.2.2 Germany

In Germany, Europe's most significant automotive producer, the spare parts issue has been vigorously debated in the Bundestag, most recently on the occasion of a proposed amendment of the German design law then in effect. Adoption of a repair clause was supported primarily by the interests surrounding the spare parts industry.[94] Several versions were discussed, including excepting from design protection spare parts used to restore a complex product to its original appearance, maintaining the existing design protection, while providing third party use under a remunerated license, limiting design protection to a pre-determined term.[95] Because consensus could not be obtained and because the European Commission was

---

[90]Grabrucker 2015, p. 20.

[91]Grabrucker 2015, p. 20.

[92]See Würtenberger and Loschelder 2015, p. 349; also Corte Suprema di Cassazione, 18 November 2011, 47081 Nr. 21. This is confirmed by Recital 7 of the Trademark Directive.

[93]See also CJEU, 6 October 2015, Case C-500/14 – Ford Motor Company v. Wheeltrims s.r.l., GRUR 2016, 77, discussed infra section 5.2.2.1.

[94]Final recommendation and report of the Legislative Committee, BT-Drucks. 15/2191, p. 8; see also Wandtke and Ohst 2005, p. 98.

[95]See Final Recommendation and report of the Legislative Committee, BT-Drucks. 15/2191, p. 8. As a fourth alternative a compulsory license was proposed, see Ruhl 2010, Article 110, para 19, criticizing the solution of the German legislator, in particular with respect to a possible violation of Article 14 Design Directive; see also Wandtke and Ohst 2005, p. 98; Kur 1998, p. 980; Riehle 1997, p. 365. Thus, until implementation of the Design Directive into German law, spare parts were rarely protectable under design law, because they generally did not meet the requirements of novelty and individual character (see above). This fact favored the spare parts industry, which was thus able to commercialize such unprotectable spare parts. With implementation of the Design Directive, these requirements were lowered, so that original manufacturers were able to obtain design protection for spare parts and thus prevent spare parts manufacturers from copying them. The concern was expressed that maintaining the lack of regulation for must match parts would in the end not liberalize, but instead restrict the market, so that inaction by the German legislator would be in violation of Article 14 Design Directive (Riehle 1997, p. 365; Riehle 1993, p. 55). This argument was countered on several grounds: first, the assumption is questionable that implementation of the Design Directive will, in practice, palpably lower the protection threshold; second, it would be hardly possible to prevent this result, either from a legal or a practical standpoint, in particular a "gap" in protection thresholds for spare parts compared to all other products must be; third, from the perspective of the German legislator it was not appropriate to create a new national regulation for must match parts, prior to the European Commission's expected harmonizing by their Proposal for a Directive amending Directive 98/71/EC, 2004 (Kur 1998, pp. 980 et seq.; see also Kur 2002, p. 669).

expected to address the spare parts issue in any event, the German legislator felt that creating a new law would not be appropriate.[96] Following negotiations with the automobile industry, the German government therefore agreed not to pursue these legislative proposals in exchange for an agreement by the automotive manufacturers not to restrict competition and to forego assertion of their design rights in spare parts used for repair.[97] This non-assertion agreement continues to be in effect.

4.2.2 *The Community Design Regulation*

While efforts relating to the adoption of the repair clause in the Design Directive and at national level were ongoing, at European level a repair clause was enacted as Article 110(1) CDR. It exempts from design protection spare parts, which constitute a component part of a complex product, when used for the purpose of the repair, so as to restore the product's original appearance. The clause operates to shield producers of spare parts from liability for infringement of a registered EU Community design. Its application is limited neither by industry nor by the nature or type of parts.[98]

To qualify for the exemption, a component part must be used in a way that is *objectively necessary* to effect a repair that restores the complex product's original appearance.[99] Objective necessity is met if the component's design corresponds to that of the original, so that after installation, the overall appearance of the complex product remains unchanged from its original state.[100] The objectiveness requirement is not expressly set forth in Article 110(1) CDR;[101] rather it is implicit in the concept of a "complex product". Absent objective necessity that the part match the appearance of the complex product, it would not qualify as a must match part, as

[96]Cf. German government's proposed law, BT-Drucks. 15/1075, p. 66.

[97]The German government's explanation relating to the proposed law, BR-Drucks. 238/03, p. 2; see also Beldiman and Blanke-Roeser 2015, p. 917. Possibly it was easy for the manufacturers to agree, as such rights had not been enforceable until then, cf. Wandtke and Ohst 2005, p. 98; in opposition the statement of the Bundestag opposition parties at the time, see final recommendation and report of the Legislative Committee, BT-Drucks. 15/2191, p. 8; also critical Kur 2016, p. 22 ("lediglich politische Wirkung"). See also § 73(1) DesignG. Because of its weaknesses, it should not serve as a role model in the decision on implementing a repair clause into European or other laws, or not.

[98]"(P)rotection as a Community design shall not exist for a design which constitutes a component part of a complex product used within the meaning of Article 19(1) for the purpose of the repair of that complex product so as to restore its original appearance".

[99]OLG Stuttgart, 11 September 2014, 2 U 46/14, GRUR 2015, 380.

[100]OLG Stuttgart, 11 September 2014, 2 U 46/14, GRUR 2015, 380.

[101]The article is worded subjectively in that it requires a subjective intent to use the part for the specified repairs.

envisioned by the legislator.[102] Separately, the purpose of the part's use must meet the subjective requirement that it serve for repair of a complex product. This intertwines with the objective necessity requirement that only matching parts qualify.[103] Non-matching parts would not satisfy the subjective premise that the purpose of use is the repair of a complex product.[104] Interpretation of this requirement thus boils down to the fact that a part that would alter the complex product's overall appearance would not be marketable.[105]

The repair clause has been asserted in recent lawsuits throughout various member countries of the EU, resulting in interpretations by several courts, which will be discussed in Chapter 5.

4.3 Legislative Efforts to Enact a Repair Clause in the U.S.

4.3.1 Stakeholder Positions

In the U.S., the legislative debate proceeds largely along the same arguments as in the EU, with an equal amount of polarity between the stakeholders' positions.

Enactment of a repair clause is supported primarily by the U.S. spare parts industry which has gained considerable economic significance and has organized effective lobby groups and alliances for purposes of representation in front of legislative bodies.[106] Their interests are aligned with and strengthened by the powerful insurance industry and opposed to the positions taken by the automotive manufacturers who are equally effectively represented in front of legislators. Compared to the EU, the discourse in the U.S. tends to rely less on legal doctrine and more on economic and business arguments. The proponents of the clause essentially argue that the steadily increasing use of design patents causes OEMs to control the spare parts aftermarket and that a repair clause is necessary to allow competition in the aftermarket and reduce prices.[107] Opponents of the clause assert that enactment of a repair clause amounts to an unconstitutional "taking" under the 5th and 14th Amendments by depriving design patent owners of the benefits of their property rights in design patents.[108] A repair clause would further ignore the

---

[102]Cf. OLG Stuttgart, 11 September 2014, 2 U 46/14, GRUR 2015, 380: "economic constraint"; Ruhl 2010, Article 110, para 29; Pentheroudakis 2002, p. 685: "economically dictated shapes".

[103]Auler, in: Büscher et al. 2015, GGV, Article 110, para 3; Ruhl 2010, Article 110, para 30.

[104]See also Ruhl 2010, Article 110, para 23, who states that the expressly provided subjective requirements of Article 110(1) CDR are balanced by the tacitly assumed objective requirements.

[105]OLG Stuttgart, 11 September 2014, 2 U 46/14, GRUR 2015, 380; Ruhl 2010, Article 110, para 29.

[106]Some of the most representatives are: Automotive Body Parts Association, Keep Auto Parts Affordable, PCI Aftermarket Parts.

[107]Efroni 2012.

[108]See AIPLA Resolution 2015.

historical balancing of the "repair versus reconstruction" distinction, a doctrine which is best left to the judiciary.[109] Finally, a repair clause would set a dangerous precedent in allowing exceptions in situations where IP protection leads to increased prices. On the safety front, automobile manufacturers also assert that independent spare parts may perform less well than the original parts from a quality perspective and that they present a safety risk subjecting auto manufacturers to safety, product liability and warranty issues.[110]

4.3.2 Proposed Bills

These diametrically opposed positions and the intensive lobbying efforts on both sides of the debate make legislative action difficult. This explains why the numerous attempts to introduce into U.S. Congress a legislative proposal in the nature of the EU repair clause have so far remained unsuccessful. The first bills were introduced by California Congresswoman Zoe Lofgren as early as 2007, and have since then been re-introduced on an almost yearly basis.[111] The early versions merely exempted from infringement component parts used for purposes of repair to restore the original appearance.[112] This broad version would have applied to infringements of utility patents as well as of design patents. It would have applied to any type of repairs and would have covered industries other than the automotive industry.

In the course of the years, the terms of the proposed bills were progressively refined and narrowed.[113] The version introduced in the 2015–2016 congressional session, titled "Promoting Automotive Repair, Trade and Sales Act" (PARTS

[109]For example, in cases such as Aro Mfg. Co. v. Convertible Top Replacement Co., 365 U.S. 336, 342, 81 S. Ct. 599, 602, 5 L. Ed. 2d 592, 128, 128 U.S.P.Q. 354 (1961) (Aro I); see also AIPLA Resolution 2015.

[110]Cf. Automotive Service Association (ASA), letter to Howard Coble (R-N.C.), chairman, and Mel Watt (D-N.C.), ranking member of the House Judiciary Committee's Subcommittee on Intellectual Property, Competition and the Internet, asking that they oppose H.R. 1663.

[111]The more recent versions have been co-sponsored by Congressman Darrel Issa, also from California.

[112]H.R.5638 — 110th Congress (2007–2008): "Section 271 of title 35, United States Code, is amended by adding at the end the following new subsection: '(j) It shall not be an act of infringement to make, use, offer to sell, or sell within the United States or import into the United States any article of manufacture that itself constitutes a component part of another article of manufacture, if the sole purpose of the component part is for the repair of the article of manufacture of which it is a part so as to restore its original appearance'".

[113]See prior Lofgren bill 2009 – 3059: "'(j) It shall not be an act of infringement of any design patent to make, use, offer to sell, or sell within the United States or import into the United States any article of manufacture that itself constitutes a component part of another article of manufacture, if the sole purpose of the component part is for the repair of the article of manufacture of which it is a part so as to restore its original appearance'".

Act),[114] would operate as exception from design patent infringement for certain external component parts of automobiles.[115] Suggested examples of covered component parts such as hoods, fenders, tail lights, and side mirrors, are expressly listed in the bill. To qualify, parts must be "similar or the same in appearance" and be intended to repair a motor vehicle and restore its appearance as originally manufactured. Unlike Article 110(1) CDR, the bill grants design patent owners a period of exclusivity of 30 months for purpose of sales, following which the design patent can no longer be asserted. On the other hand, non-sale activities, such as manufacture, testing, marketing, and pre-sale activities are permitted at any time without infringing upon design patents. The legislation would not impact automotive manufacturers' rights to enforce design patents on parts for their full duration against other car companies.[116] In short, the effect of this bill would be that alternative suppliers could offer their own generic parts for sale after the expiration of 30 months after the product is first placed on the market. Should Congress not take action in the course of the congressional session, it is probable that an identical or very similar one will be introduced in the next session.

---

[114]In February 2015, members of the House and Senate each re-introduced the "Promoting Automotive Repair, Trade and Sales Act", known as the "PARTS Act" (identical versions: H. R.1057, Congressional session 2015–2016, co-sponsored by Representatives Darrell Issa (R-CA) and Zoe Lofgren (D-CA). Similar bills have been introduced in Congress in the 2012, 2013, 2015 sessions, as well as prior to that.

[115]The bill proposed to amend 35 U.S.C. § 271 reads '(j)(1) With respect to a design patent that claims a component part of a motor vehicle as originally manufactured – (A) it shall not be an act of infringement of such design patent to make or offer to sell within the United States, or import into the United States, any article of manufacture that is similar or the same in appearance to the component part that is claimed in such design patent if the purpose of such article of manufacture is for the repair of a motor vehicle so as to restore such vehicle to its appearance as originally manufactured; and (B) after the expiration of a period of 30 months beginning on the first day on which any such component part is first offered to the public for sale as part of a motor vehicle in any country, it shall not be an act of infringement of such design patent to use or sell within the United States any article of manufacture that is similar or the same in appearance to the component part that is claimed in such design patent if the purpose of such article of manufacture is for the repair of a motor vehicle so as to restore such vehicle to its appearance as originally manufactured. "(2) For purposes of this subsection – "(A) the term 'component part' – "(i) means a component part of the exterior of a motor vehicle only, such as a hood, fender, tail light, side mirror, or quarter panel; and "(ii) does not include an inflatable restraint system or other component part located in the interior of a motor vehicle; "(B) the term 'motor vehicle' has the meaning given that term in Section 32101(7) of title 49; "(C) the term 'make' includes any testing of an article of manufacture; and "(D) the term 'offer to sell' includes any marketing of an article of manufacture to prospective purchasers or users and any pre-sale distribution of the article of manufacture".

[116]For text, see PARTS Act.

4.4 The Repair Clause in Australia

A statutory repair clause was enacted in Australia in 2003.[117] Its general tenor is similar to that of Article 110(1) CDR. The clause provides a complete defense against infringement where a component part of a complex product embodying a registered design or a design substantially similar to the registered design, is used for the purpose of repair of a complex product, so as to restore its overall appearance in whole or in part.[118] The provision is not limited by industry. Unlike Article 110(1) CDR the Australian law defines repair as restoring or replacing a "damaged or decayed" component.[119] The clause is therefore more narrowly drawn than Article 110(1) CDR and would, for instance, not apply to replacement products such as toner cartridges.[120] Furthermore certain specifics are provided for, which do not appear in the CDR repair clause. For instance, the burden of proving that the infringer knew or should have known that the use is not for purposes of repair to restore the overall appearance, is expressly placed on the owner of the design.[121] This is a critical issue for application of the clause. While it is possible for the owner of the design to satisfy the burden by an inference based on e.g. the manner in which the product is distributed, say sets of four wheel caps, this solution places the burden on the party which is not in possession of the necessary information. Alternatively, the burden could be placed on the seller/repairer, the party who is in control of the information. The seller/repairer would be in a position to furnish evidence, say by declarations, etc. and thus in a better position to meet the evidentiary burden.[122]

[117]Design Act, 2003, c. 6, § 72(1); see full text in Annex; see also Hawker 2010, pp. 8 et seq.

[118]Wanstrath 2010, pp. 409 et seq.; see also Hawker 2010, pp. 8 et seq.

[119]Not unlike Article 110(1) CDR, Section 72(1) of the Australian Design Act creates a defense against infringement of the design right to a component part of a complex product, if use of the infringing product is for purposes of repairing in order to restore the original appearance of the product. Specifically it prohibits use of products identical or substantially similar in overall impression to a registered design, for the purpose of the repair of the complex product, so as to restore the overall appearance of a complex product. Unlike Article 110(1) CDR, additional specifics are provided for in (Section 72(1) of the Australian Design Act) including (1) repair includes restoring, replacing or maintaining a decayed or damaged part, as well as necessary replacements incidental to a repair; (2) the burden of proving that the infringer knew or should have known that the use is not for purposes of repair to restore the overall appearance, is expressly placed on the owner of the design; (3) the effect of the repair is that the appearance immediately after the repair does not materially differ from the original; (4) any difference between the original and repaired product immediately after the repair is solely attributable to the fact that only part of the complex product has been repaired. The full text of the Act is available at: https://www.legislation.gov.au/Details/C2013C00156.

[120]However, these might also be excluded based on the argument that their replacement does not restore the overall appearance, as the cartridges are normally not visible.

[121]Australian Design Act, 2003, c. 6, Section 72(1)(2).

[122]A similar issue was raised in BGH, 2 June 2016, I ZR 226/14 (OLG Stuttgart), GRUR Int 2016, 1057 – Kraftfahrzeugfelgen, see infra section 5.2.3.

The impact of the Australian statutory repair clause was reviewed in a report issued in 2005.[123] The report found that all relevant issues had already been considered in the course of the legislative debates leading up to the adoption of the law in 2003[124] and that, so far, no further concrete information had emerged. The impact of the repair clause on industry and consumers therefore remained unclear. In part the lack of further data was due to the long lead-time between design registration and the time that the products appear on the market, and to the fact that the provisions have yet to be tested in the Australian courts.[125] Accordingly, the government recommended that the law remains unchanged.[126]

This concludes the review of the repair clause from a legislative perspective, discussing statutory enactments and attempts to achieve such enactments, such as in the U.S. The following sections will focus on how the repair clause has been handled by the courts.

References

American Intellectual Property Law Association (AIPLA) (2015) Draft resolution on the PARTS bill (AIPLA Resolution). http://www.aipla.org/committees/committee_pages/Industrial-Designs/Committee%20Documents/Forms/AllItems.aspx

Autorité de la concurrence, Avis n° 12-A-21 du 8 octobre 2012 relatif au fonctionnement concurrentiel des secteurs de la réparation et de l'entretien de véhicules et de la fabrication et de la distribution de pièces de rechange (The French Competition Authority's report of 2012). http://www.autoritedelaconcurrence.fr/pdf/avis/12a21.pdf

Beier FK (1994) Der Musterschutz von Ersatzteilen in den Vorschlägen für ein Europäisches Musterrecht. GRUR Int 716–732

Beldiman D, Blanke-Roeser C (2015) European design law: considerations relating to protection of spare parts for restoring a complex product's original appearance. IIC 46:915–919

Berns E (2013) Marktmissbrauch auf Ersatzteilmärkten im deutschen, europäischen und US-Amerikanischen Markt [Diss.]. EUL Verlag, Lohmar

Büscher W, Dittmer S, Schiwy P (eds) (2015) Gewerblicher Rechtsschutz, Urheberrecht, Medienrecht, 3rd edn. Carl Heymanns Verlag, Berlin

Cornels J (2015) Die Schranken des Designrechts. V & R unipress, Göttingen

Davidson M, Monotti A, Wiseman L (2016) Australian intellectual property law, 3rd edn. Cambridge University Press, Cambridge (UK)

Drexl J, Hilty R, Kur, A (2005) Designschutz für Ersatzteile – Der Kommissionsvorschlag zur Einführung einer Reparaturklausel. GRUR Int 449–457

Ebert-Weidenfeller A (2014) Reparaturklausel des Art. 110 GGV bei Verletzung deutscher Geschmacksmusterrechte nicht anwendbar, GRUR-Prax 56

[123] IP Australia, Review of the Spare Parts Provision in the Designs Act 2003 (2005); for details, see Davidson et al. 2016, p. 378.

[124] For details, see Davidson et al. 2016, pp. 360 et seq.

[125] Wanstrath 2010, pp. 409 et seq.

[126] See Wanstrath 2010, pp. 409 et seq.

Efroni Z (2012) US parts bill threatens design protection of OEM parts. http://www.automotiveworld.com/analysis/suppliers-analysis/94694-us-parts-bill-threatens-design-protection-of-oem-parts/
Eichmann H (1997) Kein Geschmacksmusterschutz für must-match-Teile?, GRUR Int 595–609
Europe Economics, the economic review of industrial rights in Europe—final report, MARKT/2013/064//D2ST/OP, January 2015, commissioned by the European Commission). http://ec.europa.eu/growth/tools-databases/newsroom/cf/itemdetail.cfm?item_id=8844
European Commission (2004) Proposal for a Directive of the European Parliament and of the Council amending directive 98/71/EC on the legal protection of designs (COM(2004) 582 final —2004/0203 (COD). http://www.eumonitor.eu/9353000/1/j9vvik7m1c3gyxp/vi7jgsymsdz3
European Policy Evaluation Consortium, Impact assessment of possible options to liberalise the aftermarket in spare parts—final report to DG internal market (EPEC Report), Paris, 18 Nov 2003. http://oami.europa.eu/en/office/pdf/spare.pdf
Fittante A (2001) Die Umsetzung der Richtlinie 98/71/EG über den rechtlichen Schutze von Mustern und Modellen: Die vom italienischen Ministerrat am 2. 2. 2001 beschlossene Gesetzesverordnung Nr. 95. The European Legal Forum 532–538
Franzosi M (ed) (1996) European design protection. Commentary to directive and regulation proposals. Kluwer Law International, The Hague, London, Boston
Grabrucker M (2015) Stellungnahme zum Vorlageersuchen an den Europäischen Gerichtshof (C-500/14) des italienischen Tribunale di Torino sezione specializata in material di imprese (Anche quale Tribunale dei marchi comunitari) betreffend ein Verfahren über die Verletzung von Markenrechten der Fa Ford Motor Company gegen Wheeltrims s.r.l. (Ersatzteilehersteller). To be published
Guizzardi S (2005) Reparaturklauseln und Eintragungsfähigkeit von Ersatzteilen in Italien – Probleme und Perspektiven. GRUR Int 299–304
Hackbarth R (2015) Reparaturklausel im Markenrecht nicht anwendbar. GRUR-Prax 481
Hartwig H (2015) Repair clause defence under European design law—a valid defence against trademark infringement? In: Marques, Class 99. http://www.marques.org/class99/Default.asp?XID=BHA618
Hartwig H (2016) Spare parts under European design and trade mark law. GRUR Int 102–108
Hawker N (2010) The automobile aftermarket: crash parts, design patents, and the escape from competition. http://www.antitrustinstitute.org/sites/default/files/aai%20collision%20repair%20parts%20commentary_032220101350.pdf
Klawitter C (2015) "Italienischer Torpedo" und Reparaturklausel in Art. 110 GGV. GRUR-Prax 466
Kroher J (1993) EG-Geschmacksmusterschutz für Kraftfahrzeug-Ersatzteile. GRUR Int 457–465
Kur A (1998) Europäische Geschmacksmusterrechtsrichtlinie verabschiedet. GRUR Int 977–982
Kur A (1999) „Freeze plus" melts the ice—observations on the European design directive. IIC 30:620–632
Kur A (2002) Die Auswirkungen des neuen Geschmacksmusterrechts auf die Praxis. GRUR 661–670
Kur A (2016) Ersatzteilfreiheit zwischen Marken- und Designrecht. GRUR 20–30
Landes WM, Posner RA (2003) The Economic Structure of Intellectual Property Law. Harvard University Press, Cambridge, MA (USA)
Llewelyn D, Barresi V (2009) Right holders' control over repair and conditioning. In: Heath C, Sanders AK (eds) Spares, repairs, and intellectual property rights. Kluwer Law International, Alphen aan den Rijn (Netherlands), pp 1–19
Oake RG (2011) Design patent perspective: understanding functionality in design patent law: Part I. In: Intellectual property today. www.designpatentschool.com/assets/Oake_OCT11%20V2.pdf
Pentheroudakis C (2002) Die Umsetzung der Richtlinie 98/71/EG über den rechtlichen Schutz von Mustern und Modellen in den EU-Mitgliedstaaten. GRUR Int 668–682
Pierson M, Ahrens T, Fischer K (2010) Recht des geistigen Eigentums: Patente, Marken, Urheberrecht, Design, 2nd edn. Vahlen, München

Riehle G (1990) Geschmacksmusterschutz für Kraftfahrzeugteile – Barriere für einen europäischen Binnenmarkt? In: Bauer JF, Hopt KJ, Mailänder KP (eds) Festschrift für Ernst Steindorff zum 70. Geburtstag am 13. März 1990. De Gruyter, Berlin, pp 911–936

Riehle G (1993) EG-Geschmacksmusterschutz und Kraftfahrzeug-Ersatzteile. GRUR Int 49–70

Riehle G (1996) Das künftige europäische Musterrecht und die „Ersatzteilfrage", EWS Beilage, 1 zu Heft 1996

Riehle G (1997) Kapituliert Europa vor der Ersatzteilfrage? „Free-for-all" und das künftige europäische Musterrecht. *EWS* 361–366

Riehle G (2011a) Immaterialgüterschutz in Sekundärmärkten. Die „Ersatzteilfrage" – Präzedenz für einen Paradigmenwechsel? In: Bechtold S, Jickeli J, Rohe M (eds), Recht, Ordnung und Wettbewerb. Festschrift zum 70. Geburtstag von Wernhard Möschel. Nomos, Baden-Baden, pp 1075–1096

Riehle G (2011b) Why a few member states should no longer block the liberalisation of the vehicle spare parts market. ECAR—analysis of the legal, economic and political status of the "Repairs Clause". Proposal of the European Commission to amend "Design"—Directive 98/71/EC [COM(2004) 582 final 14.9.2004—2004/2003 (COD)]. http://www.ecar-eu.com/documents/ECARExtendedPosition-Analysis_1July2011doc.pdf

Ruhl O (2010) Gemeinschaftsgeschmacksmuster. Kommentar, 2nd edn. Carl Heymanns Verlag, Köln

Saidman PJ (2009) Functionality and design patent validity and infringement. J Pat Trademark Off Soc 91:313

Schriever E, Schwörer C, Wildt D (2014) Deutscher Anwaltverein Brüssel, Büro Brüssel—Europa im Überblick. http://anwaltverein.de/de/newsroom/europa-im-ueberblick-20-14?file=files/anwaltverein.de/downloads/newsroom/europa_im_ueberblick/ 2014/EiUe-20-14final.pdf

Senftleben M (2010) The international three-step test. A model provision for EC fair use Legislation. JIPITEC 1:67–82

Straus J (2005) Ende des Geschmacksmusterschutzes für Ersatzteile in Europa?. Das Mandat der Kommission und seine zweifelhafte Ausführung. GRUR Int, Vorgeschlagene Änderungen der EU-Richtlinie, pp 965–979

Wahlen HJ, Kloster M (2012) Warum die Bundesregierung die überfällige Liberalisierung des Auto-Ersatzteile-Marktes *jetzt* zügig verwirklichen sollte. Positionspapier Gesamtverband Autoteile-Handel e.V. http://www.gva.de/website/files/dokumente/GVA_Positionspapier_4–2012.pdf

Wandtke AA, Ohst C (2005) Zur Reform des deutschen Geschmacksmustergesetzes. GRUR Int 91–102

Wanstrath KY (2010) Will it achieve its goal or hurt an already struggling industry? Depaul J Art Technol Intell Prop Law 20:409–448

Würtenberger G, Loschelder M (2015) Stellungnahme des GRUR-Fachausschusses für Designrecht unter Mitwirkung von Mitgliedern des Fachausschusses für Wettbewerbs- und Markenrecht zum Vorabentscheidungsersuchen des Tribunale Ordinario di Torino „Ford Motor Company/Wheeltrims SRL" (Rs. C-500/14). GRUR 348–351

Chapter 5
Jurisprudence Relating to Design Law in the Context of Spare Parts

5.1 Introduction

If the legislative front has been relatively slow in resolving the question of protectability of spare parts for repair purposes, activity on the judicial front has considerably intensified in recent years.

A comparison of lawsuits across jurisdictions yields an interesting picture. Even though the underlying legal conditions may differ, the global nature of the issues makes for numerous commonalities: most suits involve similarly placed stakeholders (often the very same players) and virtually similar factual scenarios. Most typically, a dispute will be triggered by an action brought by a manufacturer, usually automotive, for infringement of its design right in one or more visible spare parts. The possible parts at issue in these suits are quite limited in number.[1] The defendant is normally an independent spare parts producer or distributor. However, notably, in the U.S., some cases have departed from this pattern, in that spare parts manufacturers have also filed suits, offensively asserting invalidity of the parts at issue.[2]

In terms of the topics adjudicated by courts, two noticeable patterns have emerged: in countries in which a repair clause has been adopted, cases have tended to focus on developing criteria for the interpretation of the repair clause, its relation to other laws, the fine-tuning of some of the requirements and defining the scope of the clause with respect to the parts covered. In countries which have not adopted the clause, there is pressure on courts to reach individually crafted solutions with an effect comparable to that of the repair clause. This is primarily done by seeking to invalidate the design protection of spare parts, but may also be based on the application of judicially created doctrines that exempt spare part use from infringement liability, such as the repair clause. Recent years have displayed a trend

[1]The vast majority of the cases have involved wheel caps and rims.
[2]Automotive Body Parts Association v. Ford Global Technologies, filed in the ED Tex, 2013.

D. Beldiman and C. Blanke-Roeser, *An International Perspective on Design Protection of Visible Spare Parts*, SpringerBriefs in Law,
DOI 10.1007/978-3-319-54060-3_5

towards enhanced litigation in this field, that evidences a "bubbling up" of the mounting underlying tensions among stakeholders.

The following will examine court decisions from the EU, U.S. and South Africa which have proved most relevant in shaping the parameters of the repair clause doctrine in the spare parts field.

5.2 EU Jurisprudence

5.2.1 Judicially Developed Doctrines that Limit Design Protection Prior to the Adoption of Article 110(1) CDR

European jurisprudence relating to spare parts predates the discussion about EU design law in the late 1990s and indeed the enactment of the first UK design law in 1988.[3] This is why in one of the first decisions that shaped the field, plaintiff British Leyland, in *British Leyland Motor Corp. v. Armstrong Patents Co.*,[4] had to rely on copyright law in order to enforce its rights in an automobile exhaust pipe.

The case might have been resolved by holding that copyright law merely protects the blueprint for the spare part at issue, and not its reproduction as a three-dimensional article, as claimed by the plaintiff.[5] However, the House of Lords took the opportunity to weigh in on the policy aspects of the spare parts debate. Lord Bridge pointed out that consumers have "an inherent right to repair their cars in the most economical way possible".[6] This right can be of value, only if the freedom exists to acquire the spare parts necessary in an unrestricted market.[7] For that reason auto manufacturers are "not entitled to derogate from or interfere with (a consumer's right to repair) by asserting their copyright (…) against a manufacturer of parts used for repair".[8]

[3]See generally Fitzpatrick 1989, pp. 233 et seq. In British Leyland Motor Corp. v. Armstrong Patents Co. [1986] UKHL 7 (27 February 1986), the House of Lords declared the doctrine of non-derogation from grants, which is similar to the doctrine of repair and reconstruction, as it exists in U.S. law. It means that the repair of a patented product is permitted, but not its reconstruction. For civil law countries, § 9 nr. 1 of the German Patent Act (PatG) in German patent law, naming producing but not repairing as an prohibited act, can serve as an example.

[4]British Leyland Motor Corp. v. Armstrong Patents Co. [1986] UKHL 7 (27 February 1986).

[5]See Fitzpatrick 1989, pp. 233 et seq.

[6]British Leyland Motor Corp. v. Armstrong Patents Co. [1986] UKHL 7 (27 February 1986).

[7]British Leyland Motor Corp. v. Armstrong Patents Co. [1986] UKHL 7 (27 February 1986).

[8]They "already enjoyed the primary benefit which their copyright protects" in the original sale of the car. To enforce auto manufacturers' copyrights and to thereby maintain their monopolies in the supply of repair parts is "to detract from the (car) owner's rights and, at least potentially, the value of their cars" (British Leyland Motor Corp. v. Armstrong Patents Co. [1986] UKHL 7 (27 February 1986); see Fitzpatrick 1989, p. 249).

This thinking, combined with the ongoing spare parts debate, was most likely influential in the further development of UK design law. Shortly after issuance of the British Leyland decision, the UK Copyright, Designs and Patents Act of 1988 was passed, which provided for a new design right protecting the shape or configuration of three-dimensional articles and excluding both must fit and must match parts from protection.[9] Following the passage of the Design Directive, provisions substantially similar to Article 110(1) CDR were implemented into the UK design law (sec. 213). British design law at present includes a must fit provision with excludes functional designs, and a must match exception,[10] which precludes the design right from extending to design features which are "dependent upon appearance of another article of which the article is intended by the designer to form an integral part".[11]

Under certain circumstances, an exhaustion type argument may also operate as a judicially created exception to design protection. Under German design law as well as those of other EU member states, rightholders have the right to prohibit the unauthorized use of a registered design.[12] "Use" covers several actions, such as the making of a product in which the design is incorporated.[13] Not expressly mentioned is the "repair" of a product, which, as a commonly performed action to maintain the product's function, does not fall under the term making.[14] The reason for this is the doctrine of exhaustion, under § 48 DesignG.[15] This doctrine provides that once a product has been placed on the market in a EU member state, by or with the consent of the rightholder, the rightholder may no longer prohibit certain actions,[16] including the repair of a product.[17] However, it must be assessed on a case-by-case basis whether the repair actually preserves the product in question, or whether it is in fact a remanufacture of the product, as "making" cannot be object of

[9]See Copyright, Designs and Patents Act 1988 (c 48). For details on sec. 213, see Stoate and Smith 2005, p. 17.

[10]1988 UK Copyright, Designs and Patents Act, sec. 213(3)(b)(ii).

[11]Together with the must fit exception for functional spare parts in sec. 213(3)(b)(i), these exceptions, which apparently were generated by the crash parts debate and British Leyland, make clear that very few crash parts, if any, will qualify for protection under the new British law, because they are excluded by the must match exceptions Fitzpatrick 1989, pp. 233 et seq.

[12]This prohibitive right would arise under § 38(1) s. 1 DesignG, Eichmann, in: Eichmann et al. 2015, § 38, para 9. See also the underlying Article 12(1) s. 1 Design Directive.

[13]Under § 38(1) s. 2 DesignG. See Eichmann, in: Eichmann et al. 2015, § 38, para 48 et seq. See also Article 12(1) s. 2 Design Directive.

[14]Eichmann, in: Eichmann et al. 2015, § 38, para 52.

[15]Cf. Eichmann, in: Eichmann et al. 2015, § 38, para 52. See also the underlying Article 15 Design Directive: "The rights conferred by a design right upon registration shall not extend to acts relating to a product in which a design included within the scope of protection of the design right is incorporated or to which it is applied, when the product has been put on the market in the Community by the holder of the design right or with his consent".

[16]For details on the actions covered by § 48 DesignG, see Eichmann, in: Eichmann et al. 2015, § 48, para 10 et seq. For the Italian law, see Eichmann 1997, p. 603.

[17]Eichmann, in: Eichmann et al. 2015, § 38, para 48.

exhaustion.[18] In the latter case, even if the original object is broken,[19] its replacement without the rightholder's permission remains prohibited.[20] This rule also applies to component parts of complex products which are visible during normal use,[21] including must match parts.[22] Thus, absent an expressly mandated repair clause, German design law allows a rightholder to prohibit[23] the reproduction of a spare part of a complex product, e.g. a car, which incorporates a protected design.

5.2.2 *Cases Interpreting Article 110(1) CDR*

The various attempts to shield from infringement liability by way of judicially created doctrines have been superseded by enactment of Article 110(1) CDR, which changed the dynamic of the spare parts debate in Europe. Only a few years after the clause was implemented, lawsuits began to arise under Article 110(1) CDR, giving courts the opportunity to better define its contours. A wealth of decisions, including a from the CJEU, are now available to aid in interpretation of Article 110(1) CDR from various perspectives.

5.2.2.1 Approaches to the Interpretation of Article 110(1) CDR

An initial question that concerned the courts was whether Article 110(1) CDR is to be interpreted broadly, in the spirit of enabling greater competition, or given a narrow construction.[24]

Many scholars believe that national and international law, as well as economic policy, mandate a narrow interpretation, as Article 110(1) CDR is an exceptional measure that serves policy goals in a narrow set of circumstances.[25] Several reasons support this conclusion. First, a narrow construction of Article 110(1) CDR is

[18]BGH, 27. February 2007, X ZR 38/06, GRUR 2007 – Pipettensystem; Eichmann, in: Eichmann et al. 2015, § 48, para 12; Eichmann 1996, p. 867.

[19]Eichmann 1996, p. 867.

[20]Under § 38(1) DesignG. See Eichmann 1996, p. 867.

[21]Cf. Eichmann, in: Eichmann et al. 2015, § 38, para 15.

[22]Eichmann 1996, p. 867; cf. also Eichmann 1997, pp. 603 et seq.

[23]Cf. Eichmann 1996, pp. 867 et seq.

[24]As argued by Wheeltrims in CJEU, 6 October 2015, Case C-500/14 – Ford Motor Company v. Wheeltrims s.r.l., GRUR 2016, 77, for a broad construction, and by Ford for a narrow one. But the issue of construction remains disputed. See Hilty 2005, 823.

[25]Würtenberger and Loschelder 2015, p. 349, who also attribute the nature of an exception from the location of Article 110(1) CDR. This position is questionable, as Article 110 CDR is located in Title XII, Final Provisions of the CDR, and further is merely headed "Transitional Provisions", cf. Eichmann, in: Eichmann et al. 2015, § 73, para 1 et seq.; Ruhl 2010, Article 3, para 157.

mandated by Article 26(2) TRIPS[26] which expressly refers to the implementation of "limited exceptions" to design rights.[27] This narrows the room for interpretation of Article 110(1) CDR and requires that it be construed restrictively.[28] The exceptional character of Article 110(1) CDR is further reinforced from a systemic perspective. Its placement towards the end of the CDR implies that it is intended to be a limited exception,[29] applicable in a narrow set of circumstances,[30] and not to exclude spare parts from protection generally. Economic policy considerations also support a narrow construction, as uncompensated limitations of protection are only justified in exceptional situations.[31] Finally, Article 17(2) of the Charter of the Fundamental Rights of the European Union offers protection to intellectual property rights. This fact also speaks against a broad construction of regulations that limit protection of such rights.[32]

Consistent with these principles of narrow interpretation, each individual set of facts must be examined as to whether it falls within the purview of Article 110(1) CDR. The exclusion of Article 110(1) CDR is justified by the fact that no price competition exists in the secondary market for spare parts.[33] This assumption however, applies only as to those spare parts whose identical appearance is objectively necessary, as only those spare parts are capable of giving rise to

[26]Article 26(2) TRIPS: "Members may provide limited exceptions to the protection of industrial designs, provided that such exceptions do not unreasonably conflict with the normal exploitation of protected industrial designs and do not unreasonably prejudice the legitimate interests of the owner of the protected design, taking account of the legitimate interests of third parties".

[27]Hartwig 2016, p. 105; Pentheroudakis 2002, p. 669: Further international conventions which contain minimum standards for protection of design are the Berne Convention for Protection of Literary and Artistic Works (Article 2 s. 7), the Paris Convention (Article 5quinquies), the Hague Convention for International Deposit of industrial designs, as well as the Locarno Agreement for Establishment of an international classification of industrial designs, the latter two not being binding on all member states. In particular the Berne and Paris Conventions do not provide for a specific type of protection; instead they are based on general principles of intellectual property, cf. Beier 1994, p. 727 et seq., according to whom the repair clause violates the Berne and Paris Conventions (the latter in its 1993 version), as well as the Hague Convention. For further details, see Ruhl 2010, Article 110, para 15 et seq.; Drexl et al. 2005, p. 455. See Straus 2005, pp. 970 et seq. and Beier 1994, p. 728, for doubts on the conformity of a repair clause with Article 26(2) TRIPS.

[28]OLG Stuttgart, 11 September 2014, 2 U 46/14, GRUR 2015, 380; Hartwig 2016, p. 105; Ruhl 2010, Article 110, para 20.

[29]Würtenberger and Loschelder 2015, p. 349.

[30]Würtenberger and Loschelder 2015, p. 349; Eichmann, in: Eichmann et al. 2015, § 73, para 1 et seq.; Ruhl 2010, Article 3, para 157.

[31]For instance, it cannot be based on the fact that products protected by IP rights are more expensive than others.

[32]Würtenberger and Loschelder 2015, p. 349.

[33]LG Düsseldorf, 28 November 2013, 14 c O 304/12 U, GRUR-RS 2014, 00345; LG Hamburg, 18 September 2015, 308 O 143/14, GRUR-RS 2015, 16872; to this, see Klawitter 2015, p. 466.

monopolization.[34] Therefore, a court must verify in each individual instance, whether objective necessity is present and whether an appreciable economic benefit results from applying the exception in a particular case.[35] This evaluation may limit the type of parts that fall within the scope of the repair clause. It would also preclude the effect of the repair clause for any parts whose design displays an individual stylistic function, as it allows consumers a choice independent of the appearance of the vehicle as a whole, and therefore does not pose a threat of monopolization in the secondary market.[36]

5.2.2.2 Impact of Article 110(1) CDR on Other IP Rights

Two further important clarifications to the scope of Article 110(1) CDR were made by the CJEU. In *Ford v. Wheeltrims*[37] the Court ruled on the effect of the repair clause on other laws, such as trademark law, and on whether Article 110(1) CDR gives rise to a separate subjective (positive) right assertable by the alleged infringer.[38]

Ford v. Wheeltrims arose out of an action by Ford to enforce its trademark rights against Wheeltrims, a producer of wheel covers that bore Ford trademarks and were designed to fit various sizes and models of Ford vehicles. In that claim, filed in the District Court of Turin, Italy, Wheeltrims argued that as a producer of spare parts, it should be free to operate on the market on an equal level as a producer of original spare parts and that it should be entitled to reproduce Ford wheel caps regardless of IP protection, so that the original appearance of the Ford vehicle could be restored. Article 110(1) CDR should, accordingly, allow derogation from both the Italian and European laws which prohibit reproduction of the wheel cap design and of Ford's registered trademark, respectively. Ford, on the other hand, took the position that the exception created by Article 110(1) CDR is limited to designs only and that it applies exclusively in the context of Community Design, and not to other IP rights, such as registered trademarks.

---

[34]Parts which are not "objectively necessary" to the appearance of the product can be replaced without altering the product's appearance in the user's discretion, and therefore do not give rise to a monopoly of the original manufacturer over such parts; see LG Hamburg, 18 September 2015, 308 O 143/14, GRUR-RS 2015, 16872; LG Düsseldorf, 28 November 2013, 14 c O 304/12 U, GRUR-RS 2014, 00345.

[35]The extent of such assessment must be further refined, as a court cannot, in every instance, engage in a full-fledged economic analysis of the benefits resulting from application of the clause.

[36]LG Hamburg, 18 September 2015, 308 O 143/14, GRUR-RS 2015, 16872; see also OLG Stuttgart, 11 September 2014, 2 U 46/14, GRUR 2015, 380; LG Düsseldorf, 28 November 2013, 14 c O 304/12 U, GRUR-RS 2014, 00345; Ruhl 2010, Article 110, para 16.

[37]CJEU, 6 October 2015, Case C-500/14 – Ford Motor Company v. Wheeltrims s.r.l., GRUR 2016, 77.

[38]Implicitly the decisions also answer the question whether Article 110(1) CDR creates a "subjective right" to use a third party trademark, which is consistent with a narrow interpretation of the repair clause as an exception applicable to limited circumstances only.

The District Court of Turin referred the case to the CJEU for a preliminary ruling[39] on two questions, which summarized ask (1) whether the repair clause in Article 110(1) CDR allows the production of certain parts (specifically here wheel rims) for a complex product (specifically here: automotive) in order to fully restore the complex product's appearance, given the fact that bothe the component and the complex product bear a trademark of its manufacturer and (2) whether the right to copy the design of a component includes the right to reproduce the trademark placed on the original part".[40]

In the wake of the CJEU's ruling two important points are clarified: first, the fact that the effect of the repair clause does not extend beyond design law and second, that, by implication, Article 110(1) CDR does not create a separate subjective (positive) right assertable by a spare parts producer in defense against national design laws.[41] The following will discuss the impact of the decision.

5.2.2.2.1 The Effect of Article 110(1) CDR on Trademark Law

The CJEU's narrow view of the effect of Article 110(1) CDR on other laws, specifically trademark law, does not come as a surprise. Even prior to issuance of the opinion, scholars had largely agreed that the repair clause under Article 110(1)

[39]Held that the Italian repair clause, enacted as Article 241 CPI, was effective in as a defense for Wheeltrims, Article 241 CPI guarantees a fundamental economic right for manufacturers of replacement parts to produce a perfect alternative to the original component of a complex product (CJEU, 6 October 2015, Case C-500/14 – Ford Motor Company v. Wheeltrims s.r.l., GRUR 2016, 77, para 30).

[40]The Turin court referred the following questions to the CJEU: '(1) Is it compatible with EU law to interpret Article 14 of Directive 98/71 and Article 110 of Regulation No 6/2002 as conferring on producers of replacement parts and accessories the right to use trade marks registered by third parties in order to allow the end purchaser to restore the original appearance of a complex product and, therefore, also when the proprietor of the trade mark applies the distinctive sign in question to a replacement part or accessory intended to be mounted on the complex product in such a way that it is externally visible and thus contributes to the external appearance of the complex product? (2) Is the repair clause set out in Article 14 of Directive 98/71 and Article 110 of Regulation No 6/2002 to be interpreted as constituting a subjective right for third-party producers of replacement parts and accessories and, if so, does that subjective right include the right for such third parties to use the trade mark registered by another party in respect of replacement parts and accessories, by way of derogation from the rules laid down in Regulation No 207/2009 and Directive 89/104 and, therefore, when the proprietor of the trade mark also applies the distinctive sign in question to a replacement part or accessory intended to be mounted on the complex product in such a way that it is externally visible and thus contributes to the external appearance of the complex product?'.

[41]Grabrucker 2015, pp. 7, 25: Guarantee of a positive right would provide spare parts manufacturers with a claim for use of the trademark on spare parts (according to the interpretation by Ruhl 2015, p. 757: "positive right of use"), which would entail a claim for use of the trademark against its owner and a modified burden of proof in an infringement action. For details on the judgment, see Kur 2016, pp. 20 et seq., stating that uncertainties remain with regard to questions on the relation between design and trademark law. Fezer 2016, pp. 30 et seq., in agreement with the CJEU, analyzes the judgment in the context of general considerations on the legal nature of IP.

CDR or Article 14 Design Directive[42] is not assertable as a defense against other IP laws, such as trademark law, whether expressly or by implication.[43] This conclusion was based, inter alia, on the express terms of the Design Directive. Spare parts manufacturers may not use these provisions as a basis for raising affirmative claims, nor as a defense against infringement claims.[44] The fact that Article 110(1) CDR is not intended to limit other types of IP laws,[45] further results from its placement within the Design Directive, as well as from Recital 31 of the Directive, which allows other forms of IP, such as trademarks and others, to apply to protected designs.[46] Teleological reasons also speak against the ability to assert the repair clause against other laws, such as, trademark law. Trademark law aims to prevent impairment of the trademark function, whereas design protection is meant to protect the individual designer's contribution and to encourage innovation and development of new products.[47] The effect of a repair clause on trademark law would have to be examined and endorsed by the EU legislator, for instance, by adopting an equivalent regulation for trademark law. So far, this has not been the case.[48]

The CJEU therefore endorsed this narrow approach to the construction of Article 110(1) CDR.[49] Construing Article 14 Design Directive and Article 110(1) CDR jointly, the CJEU held that Article 110(1) CDR creates no derogation from the Trademark Directive (Directive 2008/95)[50] nor the from the Trademark Regulation (Regulation No 207/2009).[51] It does not operate as a defense against a registered trademark, on the ground that use of the trademark is necessary to restore the product's

[42]Würtenberger and Loschelder 2015, p. 349.

[43]Ruhl 2015, pp. 755 et seq. (further on the requirements of an analogy, which in this case are not met); see also Beyerlein 2016b, p. 196; Würtenberger and Loschelder 2015, p. 349.

[44]Cf. Würtenberger and Loschelder 2015, p. 349.

[45]See also Grabrucker 2015, p. 7. An almost identical wording can be found in Recital 7 of the Design Directive. Furthermore, see Article 96(1) CDR respectively Article 16 Design Directive; see also Beyerlein 2016a, p. 308; Beyerlein 2016b, p. 197. The conclusion supported here is further buttressed by the existence of a separate regulation of trademark law, which is applicable to spare parts in § 23 nr. 3 MarkenG respectively Article 12(c) of the Trademark Regulation which would be subject to discrepant requirements if the repair clause were to be applied to trademark law, cf. Grabrucker 2015, p. 25.

[46]Cf. Grabrucker 2015, p. 19.

[47]See Recital 7 of the CDR; Würtenberger and Loschelder 2015, p. 350.

[48]Würtenberger and Loschelder 2015, p. 350.

[49]"(T)he answer to the questions referred is that Article 14 of Directive 98/71 and Article 110 of Regulation No 6/2002 must be interpreted as not allowing, by way of derogation from the provisions of Directive 2008/95 and Regulation No 207/2009, a manufacturer of replacement parts and accessories for motor vehicles, such as wheel covers, to affix to its products a sign identical to a trade mark registered for such products inter alia by a producer of motor vehicles, without obtaining the latter's consent, on the ground that the use thus made of that trade mark is the only way of repairing the vehicle concerned, restoring to that complex product its original appearance".

[50]Directive 2008/95/EC of the European Parliament and of the Council of 22 October 2008 to approximate the laws of the Member States relating to trade marks, L 299/25.

[51]Council Regulation (EC) No 207/2009 of 26 February 2009 on the Community trade mark, L 78/1.

original appearance. Consequently, the repair clause of Article 110(1) CDR, does not apply in the context of trademark law.[52] For this particular case, this means that the clause fails to shield Wheeltrims from liability for trademark infringement.

5.2.2.2.2 Article 110(1) CDR as a Defense Against National Design Laws

A related question is whether Article 110(1) CDR operates as a defense to national design laws. German courts have held that it does not, either expressly or by implication.[53] Express applicability of the clause is excluded by the CDR's very wording, which limits its coverage to "community designs".[54] A contrary interpretation would circumvent the intent of Article 14 Design Directive, which specifically permits member states to forego implementation of the repair clause.[55] The German legislator, for instance, has deliberately refrained from adopting a rule comparable to Article 110(1) CDR. Use of Article 110(1) CDR as a shield against German design law would undermine this legislative intent.[56] If discrepant levels of protection result within the EU, this must be accepted.[57] Finally, whether the repair clause can by implication be asserted as a defense against national design law, is a matter of national law of each jurisdiction. Allowing it to do so would contravene the principles and authorities enumerated earlier, including the CJEU's generally narrow view of the scope of Article 110(1) CDR.[58]

---

[52]See also Beyerlein 2016a, p. 308; for a more detailed analysis of the decision, see Beyerlein 2016b, pp. 196 et seq.

[53]Cf. LG Düsseldorf, 28 November 2013, 14 c O 304/12 U, GRUR-RS 2014, 00345; see also Würtenberger and Loschelder 2015, p. 349.

[54]Würtenberger and Loschelder 2015, p. 349; for the German law, see LG Düsseldorf, 28 November 2013, 14 c O 304/12 U, GRUR-RS 2014, 00345; to this Ebert-Weidenfeller 2014, p. 56.

[55]Similar Grabrucker 2015, p. 7 (in connection with applicability to trademarks).

[56]Cf. Beyerlein 2016b, p. 198; Ebert-Weidenfeller 2014, p. 56.

[57]LG Düsseldorf, 28 November 2013, 14 c O 304/12 U, GRUR-RS 2014, 00345. Headnote (as formulated by Ebert-Weidenfeller 2014, p. 56): "(I)n a claim for infringement of a German design right, the repair clause under Article 110 CDR is not applicable, either directly or by implication. The German legislator has deliberately not adopted a comparable rule. Discrepant levels of protection within the EU must be accepted".

[58]Under German law, for instance, applicability could be premised on the existence of an unintended gap in regulation, given similarly situated interests, cf. Larenz and Canaris 1995, pp. 202 et seq. Certain legislatures' failure to adopt the repair clause might indeed be construed as a gap in regulation, assuming the law is otherwise silent on the topic. However, its effect is not unplanned: any existing legal discrepancies between the CDR and the laws of such member states, are the result of deliberate action by the respective legislatures, pursuant to the latitude allowed by the European Commission, see LG Düsseldorf, 28 November 2013, 14 c O 304/12 U, GRUR-RS 2014, 00345; Ebert-Weidenfeller 2014, p. 56. Nor does such gap in regulation gain a *post factum* "unplanned" character, as a result of a future modification of Article 14 Design Directive and its ensuing implementation into national laws. In short, there is little basis upon which to argue by implication that the CDR applies to national design laws. See LG Düsseldorf, 28 November 2013, 14 c O 304/12 U, GRUR-RS 2014, 00345.

Consequently, it is likely that attempts to assert Article 110(1) CDR as a defense against national design laws in countries which have not adopted a repair clause, would be fruitless.[59]

5.2.3 *Parts Covered by Article 110(1) CDR*

Further definition of the scope of Article 110(1) CDR relates to the specific component parts covered by the clause. The clause lists no specific parts, it merely refers to "component part(s) of a complex product used (...) for the purpose of the repair of that complex product so as to restore its original appearance".

The English High Court of Justice in *BMW v. R&M*[60] and the Stuttgart Higher Regional Court[61] each examined the question whether alloy wheels fall within the purview of Article 110(1) CDR. While adopting slightly different analytical approaches, the cases reach comparable results. According to the English High Court of Justice, in order to fall within the purview of Article 110(1) CDR, a part must be "dependent" on the appearance of the complex product. In contrast, the Stuttgart court, held that a design must be "objectively necessary" to match the appearance of the complex product, and that the purpose of its use must further be restricted to repairs.

5.2.3.1 The Dependency/Objective Necessity Requirement

In deciding whether Article 110(1) CDR applies to alloy wheels, the English High Court of Justice, in BMW v. R&M, undertook an extensive analysis of the legislative history of the repair clause.[62] Reading Article 110(1) in conjunction with Recital 13 of the CDR,[63] the court concluded that "Article 110(1) only applies to component parts of a complex product 'upon whose appearance the protected

[59]LG Düsseldorf, 28 November 2013, 14 c O 304/12 U, GRUR-RS 2014, 00345.

[60]BMW AG v. Round and Metal Ltd and Philip David Gross [2012] EWHC 2099 (pat).

[61]OLG Stuttgart, 11 September 2014, 2 U 46/14, GRUR 2015, 380.

[62]BMW AG v. Round and Metal Ltd and Philip David Gross [2012] EWHC 2099 (pat): "(A) consistent aspect of the discussions throughout the process was that what was under consideration was component parts whose appearance was dependent on the appearance of the complex product. It is reasonably clear from the agreement reached by the Conciliation Committee, and in particular the nature of the consultation exercise which the European Commission agreed to undertake, that the compromise struck in the Design Directive concerned such parts. Equally, it is reasonably clear that the compromise proposals promoted by the Presidency for the wording of the Regulation were directed to the same situation" (BMW para 60).

[63]Recital 13 of the CDR states that "it is appropriate not to confer any protection as a Community design for a design which is applied to or incorporated in a product which constitutes a component part of a complex product upon whose appearance the design is dependent and which is used for the purpose of the repair of a complex product so as to restore its original appearance".

design is *dependent*'[64] (emphasis added). The purpose of the "dependency" requirement is to "ensure that the original equipment manufacturer cannot monopolize the aftermarket, where the owner of the product has *no realistic alternative* to replacing the part with one of the same design if the original part becomes damaged"[65] (emphasis added). This means that the clause does not apply to parts with respect to which a user can exercise discretion without changing the appearance of the vehicle, or that serve to upgrade the appearance of a car, because such parts are not "dependent" on the car's original appearance, within the meaning of Article 110(1) and Recital 13 of the CDR. In the case of the alloy wheels at issue in BMW v. R&M, replacing wheels intended for one design with wheels meant for another, was found to be a perfectly realistic option. Therefore it cannot be said that the alloy wheels are "dependent" on the appearance of the car. Evidence that BMW markets a range of different alloy wheels suitable for various of its models[66] and that R&M sells its replicas as suitable for multiple models,[67] indicated to the court that customers have a choice between standard wheels and upgrades or retrofits and that they are not restricted to a particular wheel design for a particular car model.[68] Based thereon, the court concluded that alloy wheels do not meet the dependency requirement and do not fall within the purview of the repair clause.

A slightly different approach to the question was adopted by German courts. In a 2014 decision, the Stuttgart Higher Regional Court construed Article 110(1) CDR to imply a requirement of "objective necessity". This means that a part qualifies for the exemption only if imitation of the original is objectively necessary in order to restore the original appearance. The court predicated this conclusion on the wording of Article 110(1) CDR. Had the legislator intended to exclude all must match parts, the exclusion would have simply referred to imitations of protected visible spare parts, when used for repair purposes.[69] The present complicated and rather unwieldy formulation requiring restoration of the original appearance, points to the fact that it was not the legislator's intent to exclude replicas of all must match parts.[70] Rather, parts which make up the visual appearance of a vehicle must be viewed as "divisible" into categories, some of which fall under the purview of the repair clause and others which do not: "Wheel rims could be one of the main areas of application of this intent".[71] This reasoning prompted the court to conclude that

[64]BMW AG v. Round and Metal Ltd and Philip David Gross [2012] EWHC 2099 (pat).

[65]BMW AG v. Round and Metal Ltd and Philip David Gross [2012] EWHC 2099 (pat).

[66]Factory standard, factory up-grade and retro-fit accessories.

[67]Additionally, in case of the MINI, the replica wheels in issue are suitable for different brands of cars.

[68]BMW AG v. Round and Metal Ltd and Philip David Gross [2012] EWHC 2099 (pat).

[69]OLG Stuttgart, 11 September 2014, 2 U 46/14, GRUR 2015, 380.

[70]OLG Stuttgart, 11 September 2014, 2 U 46/14, GRUR 2015, 380.

[71]OLG Stuttgart, 11 September 2014, 2 U 46/14, GRUR 2015, 380. The standard is that if an objective necessity to imitate exists, the vehicle is no longer marketable. Wheel rims may be equated to alloy wheels (see BMW AG v. Round and Metal Ltd and Philip David Gross [2012] EWHC 2099 (pat)).

with respect to wheel rims, there is no "objective necessity" to imitate the original design, because their design is not contingent on the appearance of the car.[72]

After having "brewed" for several years in the national courts of various countries, the issue reached the CJEU, in the form of references from the German Federal Court of Justice (BGH)[73] and the Milan Court of Appeal.[74] Both involved the scope of protection of Article 110(1) CDR.[75] The originating courts of these references held contrary positions on the matter. The German court viewed the repair clause as an exception which must be interpreted narrowly, in other words wheel rims should be excluded from the purview of Article 110(1) CDR. The Italian court on the other hand, following the opinion of the Court of Appeal of Naples,[76] gave it a broader, more general interpretation, in the spirit of liberalizing the spare parts market, consequently including wheel rims.[77] Furthermore, each of the cases separately raised the question from a slightly different angle. The referral from the German Federal Court of Justice involved an infringement suit by Porsche relating to design protected aluminum wheels. Defendants raised Article 110(1) CDR in defense and appealed a judgment by the Stuttgart Higher Regional Court in favor of Porsche. The outcome turns on the interpretation of the scope of Article 110(1) CDR. In this context the German court sought clarifications from the CJEU relating to the scope of Article 110(1) CDR, asking, in summary, the following questions:[78]

---

[72]Cf. OLG Stuttgart, 11 September 2014, 2 U 46/14, GRUR 2015, 380.

[73]BGH, 2 June 2016, I ZR 226/14 (OLG Stuttgart), GRUR Int 2016, 1057 – Kraftfahrzeugfelgen.

[74]Corte di Appello di Milano, 18 July 2016, CJEU, Case C-397/16 – Acacia s.r.l. v Pneusgarda s.r.l. (in bankruptcy), Audi AG.

[75]Ferretti 2016, p. 22; Hackbarth 2016, p. 382.

[76]Corte di Appello di Napoli, 22 October 2013, Riv. dir. ind. 2014, II, 230; for details on that, see infra section 5.2.3.2.

[77]For details, see Ferretti 2016, p. 22.

[78]The German court asks:

1. Is the application of an exception under Article 110(1) CDR limited to parts that are shape-dependent, i.e. parts whose shape is unalterably determined by the appearance of the complex product and therefore not freely selectable by the customer, such as automotive wheel rims?
2. In the event the answer to question nr. 1 is in the negative: is the application of an exception under Article 110(1) CDR limited to products that are identical in shape, color and size to the original products?
3. In the event the answer to question nr. 1 is in the negative: does the exception under Article 110(1) CDR apply to the provider of spare parts alleged to infringe an original design, only if the provider ensures that, objectively seen, its product is utilized solely for purposes of repair and not for other purposes, such as the upgrade or personalization of the complex product?
4. In the event the answer to question nr. 1 is in the affirmative: what measures must a provider of spare parts alleged to infringe an original design, undertake in order to ensure that its product is utilized solely for purposes of repair and not for other purposes, such as the upgrade or personalization of the complex product? Is it sufficient

(a) that the provider's sales materials includes a notice to the effect that the purpose of the sale is limited to repairs meant to restore the original appearance of the complex product, or

(1) whether the scope of Article 110(1) CDR is limited to form-dependent parts, whose shape is unalterably determined by the appearance of the complex product or whether it encompasses products freely selectable by the customer, such as automotive wheel rims? (2) If so limited, must the products be identical in shape, color and size to the original products? (3) If so limited, does Article 110(1) CDR apply only to products used for repair purposes or does it extend to upgrades or personalizations of the vehicle? (4) If not so limited, what measures must an independent spare parts provider take in order to ensure that they are used for repairs: is a notice in the sales materials sufficient or must the purchaser give a written declaration of use.

A narrow construction of Article 110(1) CDR would exclude parts that are not form-dependent, possibly based on the reasoning set forth in the language in Recital 13 of the CDR, also followed by English High Court in BMW v. R&M.[79] The Recital refers to designs "dependent" on the appearance of the complex product.[80] This suggests that an exemption is required only in situations where, due to the part's form-dependency, the original equipment manufacturer ends up monopolizing the aftermarket. On the other hand, where the owner of the product has a realistic alternative to replacing the part with one of the same design, there is no possibility of monopolization and the rationale for an exemption disappears.[81] If so, products which constitute upgrades and personalizations that can substitute for the original part do not fall within the purview of Article 110(1) CDR. Expressed differently, an exemption is justified only in cases in which imitation of the original is objectively necessary in order to restore the original appearance[82] and not in cases in which the parts display an individual and independent stylistic function.[83]

Under this reasoning, wheel rims, wheel caps and other parts that are freely selectable by the user, would not fall within the purview of the exclusion of Article 110(1) CDR, and design owners would be entitled to enforce design rights in parts that are not form-dependent. Were the CJEU to adopt this approach,[84] the precedential effect on the repair clause could be considerable. If the overall visual

(Footnote 78 continued)

(b) is it necessary that the provider makes the sale contingent on a written declaration by the purchaser (distributor or consumer) to the effect that the product provided will be limited to repairs?.

[79]BMW AG v. Round and Metal Ltd and Philip David Gross [2012] EWHC 2099 (pat).

[80]Recital 13 of the CDR provides that "it is appropriate not to confer any protection as a Community design for a design which is applied to or incorporated in a product which constitutes a component part of a complex product upon whose appearance the design is dependent and which is used for the purpose of the repair of a complex product so as to restore its original appearance".

[81]BMW AG v. Round and Metal Ltd and Philip David Gross [2012] EWHC 2099 (pat).

[82]OLG Stuttgart, 11 September 2014, 2 U 46/14, GRUR 2015, 380.

[83]LG Hamburg, 18 September 2015, 308 O 143/14, GRUR-RS 2015, 16872; LG Düsseldorf, 28 November 2013, 14 c O 304/12 U, GRUR-RS 2014, 00345; for details, see Kur 2016, pp. 20 et seq.

[84]The ruling would be consistent with the over 40 national court decisions in different EU member states, holding wheel rims and wheel caps are beyond the scope of the clause. See the decisions mentioned in Helsinki District Court, 19 November 2015, Nr. 15/149362, R 14/5257 – Public Prosecutor and BMW v. Hotari and Pohjola.

appearance of a vehicle is considered “divisible” depending on whether a spare part is form-dependent on the original product or not,[85] numerous other discretionary spare parts would fall outside the exemption. Given an increasing trend towards customization of vehicles, fueled by manufacturer-offered options and modifications of new cars made by authorized aftermarket retrofitters, many other visible parts, such as darkened glass, grilles, bumpers, spoilers or lights could easily fall outside the scope of the repair clause.[86]

The second reference to the CJEU came from the Milan Court of Appeal[87] and involves a suit by Audi related to infringement of its registered design by replica wheels (wheel rims) manufactured by Acacia. Audi obtained a favorable first instance decision and Acacia appealed. The Milan court directed three questions to the CJEU, which can be roughly summarized as follows[88]: (1) Does the exclusion of replica wheels from the purview of Article 110(1) CDR violate certain established principles of EU law, including the free movement of goods and services;

[85]Or, alternatively, whether it is objectively necessary to restore the product’s original appearance. See OLG Stuttgart, 11 September 2014, 2 U 46/14, GRUR 2015, 380.

[86]Retrofitters such as Barabus, Abarth and AMG are in the business of offering customized modifications of existing designs, e.g. modified body parts. See also Kroher 1993, pp. 460 et seq. As a practical matter, the effect of this judgment is that trademark rights can be used to circumvent the repair clause by affixing their trademark on individual spare parts, and claim trademark infringement in countries in which a repair clause has been enacted.

[87]Corte di Appello di Milano, 18 July 2016, CJEU, Case C-397/16 – Acacia s.r.l. v Pneusgarda s.r.l. (in bankruptcy), Audi AG.

[88]“1. Do (a) the principles of the free movement of goods and of the freedom to provide services within the internal market, (b) the principle of the effectiveness of EU competition law and of the liberalisation of the internal market, (c) the principles of *effet utile* and of the uniform application within the European Union of EU law and (d) the provisions of secondary EU law, such as Directive 98/71, 1 and in particular Article 14 thereof, Article 1 of Regulation No 461/2010 2 and UNECE Regulation Nr. 124, preclude an interpretation of Article 110 of Regulation No 6/2002, 3 which contains the repair clause, that excludes replica wheels aesthetically identical to original equipment wheels and approved on the basis of UNECE Regulation Nr. 124 from the definition of a ‘component part of a complex product’ (that complex product being a motor vehicle) for the purposes of the repair of that complex product and the restoration of its original appearance? 2. In the event that the first question is answered in the negative, do the rules on exclusive industrial rights in respect of registered designs, regard being had to the balancing of the interests referred to in the first question, preclude the application of the repair clause to replica complementary products that may be selected freely by the customer, on the basis that the repair clause is to be interpreted restrictively and may be relied upon only with respect to spare parts that come in one particular form only, that is to say, component parts the form of which has been determined in practically immutable fashion with respect to the external appearance of the complex product, to the exclusion of component parts that may be regarded as interchangeable and that may be applied freely, in accordance with the customer wishes? 3. In the event that the second question is answered in the negative, what steps must a manufacturer of replica wheels take in order to ensure the free movement of products the intended use of which is the repair of a complex product and the restoration of its original appearance?”.

(2) Does design law preclude the application of Article 110(1) CDR to goods that are freely selectable by the customer; whose form is determined immutably by the appearance of the vehicle; (3) If design law does not preclude such parts from the purview of Article 110(1) CDR, what steps must their manufacturer take to ensure their free movement?

This case raises one of the thorniest questions surrounding the repair clause and goes to the core of the lack of harmonization of spare parts laws within Europe: the tension between IP laws[89] and the free movement of goods and services.[90] European courts have long struggled with this question.[91] A holding that replica wheels fall under the repair clause would enhance that tension. In countries which have adopted the repair clause, such as Italy, sales of replica wheels within the country would be privileged (under both Article 110(1) CDR and the national repair clause). Conversely, when such wheels are sold into a country which has not adopted the repair clause, they would infringe national design laws. This outcome is problematic as it would force a court to give precedence to national IP laws over the principle of free movement of goods and services, or, alternatively, to disregard IP laws to the detriment of free movement. Both of these positions are difficult to justify. However, if the CJEU were to hold that wheel rims are excluded from the purview of Article 110(1) CDR, this dilemma could be bypassed, at least for purposes of the case at issue.[92] A holding excluding wheel rims from Article 110(1) CDR would be consistent with numerous national court decisions on this question.[93] It would also be well-grounded in the Design Regulation[94] and the principle of narrow construction of an exception to IP rights.[95] Most importantly, it would be consistent with the conceptual underpinning of Article 110(1) CDR, whose intent is to prevent monopolization in the spare parts market. Because consumer choice of

[89]Protected, inter alia, under Article 17(2) of the Charter of Fundamental Rights of the European Union.

[90]See Article 34 et seq., 45 et seq. TFEU.

[91]See e.g. CJEU, 13 July 1966, Joined cases C-56/64 and C-58/64 – Etablissements Consten SA & Grundig-Verkaufs-GmbH v Commission [1966] ECR 299; CJEU, 14 July 1981, Case C-187/80 – Merck & Co. v. Stephar BV [1981] ECR 2063.

[92]At least with respect to the case at issue, as the fundamental tension between these rights would remain.

[93]See e.g. OLG Stuttgart, 11 September 2014, 2 U 46/14, GRUR 2015, 380; LG Hamburg, 18 September 2015, 308 O 143/14, GRUR-RS 2015, 16872; LG Düsseldorf, 28 November 2013, 14 c O 304/12 U, GRUR-RS 2014, 00345; BMW AG v. Round and Metal Ltd and Philip David Gross [2012] EWHC 2099 (pat).

[94]The wording of Recital 13 of the CDR refers to "dependency".

[95]See e.g. Article 26(2) TRIPS; also supra section 5.2.2.1.

individual spare parts would introduce substitutability into the spare parts market, the need for an exemption would be obviated.

5.2.3.2 The Purpose Requirement

An alternative basis for a narrowing interpretation of Article 110(1) CDR is its purpose requirement. Article 110(1) CDR only applies to parts used for the purpose of repairing a complex product.[96] This is a subjective requirement. It would be met by evidence presented by the spare parts manufacturer showing its intention to use the part for purpose of restoration, rather than as an upgrade.[97] As such, it presents the risk of untruthful claims by the spare parts manufacturer as to the intended purpose of the parts.[98] Such claims might be countered by evidence that alloy wheels come in sets of four, are made for different models or are not "normally" used for repair.[99] In the BMW v. R&M case, evidence that most of the wheels are generally sold in sets of four (either four of the same or two pairs, front and back) and not sold singly or in pairs, indicated to the court that the proposed use was likely not a repair, but an upgrade of the vehicle's appearance.[100]

In instances where the part does not come in multiples, evidence relevant to an intent to repair could come from the manner in which the parts are distributed.[101] Parts intended for repair would be distributed by the manufacturer by an authorized distributor[102] or, alternatively, by a service provider engaged exclusively in repairs. Conversely, parts sold to private customers or resellers may not be intended for repairs. Testimony from an authorized repair service might serve to confirm whether the parts are actually being used for repair purposes.[103]

In conclusion, jurisprudence has provided a certain amount of clarification on the scope of Article 110(1) CDR. The lines of reasoning of the English High Court of Justice and the Stuttgart Higher Regional Court have been followed by numerous courts throughout Europe. Courts have consistently endorsed a narrowing interpretation of Article 110(1) CDR, and specifically held that wheel caps, rims or alloy wheels do not fall under the repair clause, because they are discretionary upgrades

[96]BMW AG v. Round and Metal Ltd and Philip David Gross [2012] EWHC 2099 (pat).

[97]Ruhl 2010, Article 110, paras 32, 34, also on the difficulties of controlling this ("Schutzumfangdelle").

[98]OLG Stuttgart, 11 September 2014, 2 U 46/14, GRUR 2015, 380.

[99]OLG Stuttgart, 11 September 2014, 2 U 46/14, GRUR 2015, 380; Hackbarth 2015, p. 481.

[100]BMW AG v. Round and Metal Ltd and Philip David Gross [2012] EWHC 2099 (pat).

[101]For details on other aspects relevant for interpretation of the clause, see Guizzardi 2005, p. 301.

[102]England and Wales High Court of Justice, 27 July 2012, [2012] EWHC 2099 [Pat]; see Brtka and Soetens 2015.

[103]Brtka and Soetens 2015; Hackbarth 2015, p. 481; Ruhl 2010, Article 110, para 32.

and not necessary to restore the original appearance. Decisions have been issued by courts in Germany,[104] Belgium,[105] Spain,[106] Denmark,[107] Finland,[108] Italy[109] and other countries.[110] The Court of Appeal of Naples, on the other hand, held wheel caps to be covered by the repair clause, applying a broad, competition-friendly construction of the clause.[111] Future CJEU decisions are expected to further flesh out the rationale for excluding certain parts based on form-dependency and the effect on free movement within the EU.

5.3 U.S. Jurisprudence

The considerable tension between players in the spare parts market is evidenced not only by the legislative efforts described above, but also by intensified litigation in recent years. Nonetheless, the U.S. has not enacted a repair clause. While judicial activity relating to design protection for spare parts remains inconclusive, courts have not been inactive.

---

[104]See also OLG Stuttgart, 11 September 2014, 2 U 46/14, GRUR 2015, 380; OLG Hamburg, 5 December 2001, 5 U 114/01, GRUR-RR 2002, 129; LG Hamburg, 18 September 2015, 308 O 143/14, GRUR-RS 2015, 16872; LG Düsseldorf, 28 November 2013, 14 c O 304/12 U, GRUR-RS 2014, 00345.

[105]Belgium: Tribunal de Commerce Francophone de Bruxelles, 16 February 2015, R.G. A/12/05787 – BMW v. SA Auto Sports Willy, SARL EUROWHEELS and THREEFACE TUNING: Wheel rims do not fall under the repair clause of Article 3.19 of the Benelux Convention on Intellectual Property (trademarks and designs).

[106]Spain: Commercial Court Nr. 2 of Alicante, 7 March 2011-544/2010. Headnote: wheel rims do not fall under the repair clause of Article 110(1) CDR.

[107]Denmark, Supreme Court, 10 March 2015, 17/2010 – BMW v. East Scandic (aluminum wheels).

[108]Finland: Helsinki District Court, 19 November 2015, Nr. 15/149362,R 14/5257—Public Prosecutor and BMW v. Hotari and Pohjola: "Aluminium rims are thus not used to restore the original appearance of the car, but in order to decorate it and to improve its overall appearance. Different rim models can be used with the same model of car, as can the same rim model be used with different models of cars. Therefore, aluminium rims cannot be within the scope of Article 110 and the provision therefore does not prevent BMW from relying on the registered designs in this case" (unofficial translation by Castrén & Snellman Attorneys Ltd, Helsinki, Finland; see also Salonen 2016).

[109]Civil Court Bologna, 17 February 2013, Nr. 4306/2011 and Nr. 9059/2011 and Tribunale di Milano, 21 February 2013, Nr. 75493/2010 (wheel rims do not fall under the repair clause of Article 110(1) CDR); Tribunale di Milano RG, 27 November 2014, Nr. 3801/2013 – Audi AG v. Acacia s.r.l; Tribunale di Milano, 19 February 2015, RG Nr. 2271/2015 – Audi AG v. Acacia s.r.l. and Fallimento Pneusgarda s.r.l. (wheel rims do not fall under the repair clause of Article 110(1) CDR or Article 241 CPI); Corte Suprema di Cassazione, 18 November 2011, Nr. 47081.

[110]See also Würtenberger and Loschelder 2015, p. 349.

[111]Corte di Appello di Napoli, 22 October 2013, Riv. dir. ind. 2014, II, 230.

5.3.1 The Repair Doctrine

Even though no repair clause has been enacted, according to some interpretations, U.S. law features a common law, non-statutory repair doctrine (or doctrine of repair and reconstruction) that entitles purchasers of a patented apparatus to repair and replace worn or broken parts.[112] Arguably, its effect could be comparable to that of a statutory repair clause.[113]

The repair doctrine is a limitation to the scope of the exhaustion doctrine.[114] The latter, also referred to as first sale doctrine, is known under U.S. law as an affirmative defense,[115] allowing purchasers to freely use and resell a product following an authorized sale.[116]

The repair doctrine's underlying reasoning is that, one who acquires a patented product, has the right to preserve the product's useful life, including the right to repair it.[117] This right to repair is strictly circumscribed by the rule that ownership does not include the right to construct an essentially new article on the template of the original:[118] reconstruction or replacement that amounts to "a second creation of the patented entity" is a right that must remain with the patentee. The dichotomy "repair"–"reconstruction" thus defines the distinction between permitted and prohibited activities with respect to patented items once they have been placed into commerce.[119] As long as the activity constitutes a repair and not a reconstruction, the repair doctrine would operate to insulate a repairer of a patented article from direct or contributory liability for patent infringement.

---

[112]Lummus Industries v. D.M. & E. Corp., 862 F.2d 267, 272 (Fed. Cir. 1988).

[113]For details on the repair doctrine, see Janis 1999, pp. 443 et seq.

[114]The exhaustion doctrine is an uncodified common law doctrine under U.S. patent law, which was first recognized in 1873 by the U.S. Supreme Court in Adams v. Burke, 84 U.S. 453 (1853). For an introduction, see Osborne 2004, pp. 646 et seq., pp. 667 et seq.

[115]On the exhaustion principle under European law, see supra section 5.2.1.

[116]See Adams v. Burke, 84 U.S. 453 (17 Wall.) 453, 21 L. Ed. 700; Osborne 2004, pp. 646 et seq.

[117]The purchaser of a patented article has the rights that an owner of personal property normally has, including the right to use it, repair, modify it, discard it, or resell it, subject only to overriding conditions of the sale (Lummus Industries v. D.M. & E. Corp., 862 F.2d 267, 272 (Fed. Cir. 1988)).

[118]See Aro Mfg. Co. v. Convertible Top Replacement Co., 365 U.S. 336, 342, 81 S. Ct. 599, 602, 5 L. Ed. 2d 592, 128, 128 U.S.P.Q. 354 (1961) (Aro I). The court permitted replacement of an automobile convertible top fabric even though the combination of the convertible top mechanism, including the fabric cover was patented. The court held the replacement to be an authorized repair of the patented product; see also Osborne 2004, pp. 667 et seq.; Janis 1999, pp. 443 et seq. But see Aro Mfg. Co. v. Convertible Top Replacement Co., 377 U.S. 476, 487, 84 S. Ct. 1526, 1532-33, 12 L. Ed. 2d 457, 141 U.S.P.Q. 681 (1964) (Aro II) where the court held there was no right to repair.

[119]With the rise of 3D printing and its potential for the aftermarket, the distinction between repair and reconstruction is likely to become even more difficult. See also Janis 1999, pp. 425 et seq. Wilbanks 2013, pp. 1147 et seq., provides an interesting legal analysis and proposals to facilitate the application of the repair doctrine.

While originally developed for use in connection with utility patents, the repair doctrine has been applied to design patents as well.[120] However, whether it extends to shield from liability for infringement a party who acquires and replaces an individually patented component part, as under German law, for example, is unsettled.[121] Courts have adopted divergent positions on this issue.

On the one hand, it has been held that the right to renew depends upon the right to "make" the invention.[122] If that right does not exist, there is no right to rebuild the combination.[123] For instance, it has been argued that the repair doctrine only applies to the repair of products actually licensed under patents because "the right of one other than the patentee, furnishing repair parts of a patented combination, can be no greater than that of the user".[124] Under this reasoning, the repair doctrine did not shield a defendant who made and sold a patented component part from liability for infringement.[125]

On the other hand, at least one case seems to suggest that the repair doctrine might apply to a patented component part.[126] A defendant's activity of selling rebuilt patented vehicle clutches, that it had assembled on a production line using parts from disassembled worn-out clutches, was held to constitute permissible repair.[127] The basis of the holding, however, is not quite clear. It is possible that the court focused on the clutch itself as the patented apparatus, rather than on its component parts. It is therefore uncertain whether the repair doctrine can serve as a defense in an infringement action in the context of spare parts for complex products.

In short, the substantive aspects of the repair clause have not truly been addressed on the merits. The repair doctrine defense thus arguably remains assertable against infringement of spare part design patents, but the outcome is uncertain.

5.3.2 *Lawsuits Relating to Spare Parts*

Litigation surrounding automotive collision spare parts in the U.S. began in the 1980s but it has intensified in the past few years. Unlike EU courts, U.S. courts have issued few final judgments and the decisions have largely eschewed the core

---

[120]Jazz Photo Corp. v. International Trade Commission, 264 F. 3d 1094 (Fed. Cir. 2001); Fuji Photo Film Co., Ltd. v. Jazz Photo Corp., 394 F.3d 1368 (Fed. Cir. 2005).

[121]For the German approach to this problem, see supra section 5.2.

[122]"Making" is meant in the sense of 35 U.S.C. § 271.

[123]Wilson v. Simpson, 50 U.S. 109, 123 (1850); see also Janis 1999, pp. 437 et seq.

[124]Aro Mfg. Co. v. Convertible Top Replacement Co., 377 U.S. 476, 487, 84 S. Ct. 1526, 1532-33, 12 L. Ed. 2d 457, 141 U.S.P.Q. 681 (1964) (Aro II).

[125]Hensley Equipment Co. v. Esco Corp., 383 F.2d 252, 260 (5th Cir. 1967), judgment amended, reh'g denied, 386 F.2d 442, 673 (5th Cir. 1967); Canon Inc. v. GCC Intern. Ltd., 450 F. Supp. 2d 243 (S.D. N.Y. 2006).

[126]Dana Corp. v. American Precision Co., 827 F.2d 755 (Fed. Cir. 1987).

[127]Dana Corp. v. American Precision Co., 827 F.2d 755 (Fed. Cir. 1987).

issues of the debate, such as the effect of the repair doctrine on component parts or the validity of design patents.[128] The reason may be in part procedural. Validity of the design patent is routinely raised in infringement actions and, depending on the facts, may prompt an early settlement. An enforcement suit thus potentially places the design patent in jeopardy. This may account for the rightholders' cautious approach to enforcement and for the dearth of decisions on the merits.

An example on point is the case in which Chrysler sought an injunction against an alleged infringer of one of its design patents.[129] The district court denied the injunction, on the ground that Chrysler had failed to show a likelihood of success on the merits, citing possible invalidity of the design patent based on anticipation and functionality.[130] The Court of Appeal affirmed the denial of the injunction, but also noted the existence of a statutory presumption of validity and the fact that doubt as to validity "does not prejudge the ultimate question on the merits of whether the patent is valid".[131] This is particularly true in notoriously complex cases of alleged functionality.[132] Eventually, no ruling on the merits was issued in the case. However, had Chrysler pursued the injunction, an invalidity ruling regarding its design patents might have been the outcome.

Nonetheless, a few infringement actions by the automotive industry have been successful. Ford brought an action with the U.S. International Trade Commission (ITC) against the Taiwanese *LKQ Corporation*, seeking to block importation into the U.S. of certain automotive parts which infringed Ford's design patents.[133] The ITC Administrative Law Judge held three of the patents invalid due to public use[134] and the remaining seven to be valid (not anticipated and not functional),

---

[128]Volvo N. Am. Corp. v. Keystone Automotive Indus., CV86-8306 CBM (GX) (complaint filed Dec. 19. 1986) for an infringement of a design patent, cited by Fitzpatrick 1989, p. 233.

[129]Chrysler Motors Corp. v. Auto Body Panels of Ohio, 719 F. Supp. 622 (S.D. Ohio 1989), aff'd 908 F.2d 951 (1990), involved design patents on a Chrysler Dodge Dakota fender.

[130]Holding based on the fact that the fender design was an "obvious combination of previous well-known design features" and that it appeared to be "designed according to functional and performance considerations as opposed to aesthetic or ornamental considerations". The court also stated that the Chrysler fender was governed by metallurgical, engineering and production considerations and determined by considerations of function or serve functional purposes (Chrysler Motors Corp. v. Auto Body Panels of Ohio, 719 F. Supp. 622 (S.D. Ohio 1989)).

[131]"The question of when the functionality of a design so permeates an article of manufacture that design patent protection is not available under the law is a complex issue and one that continues to be the subject of considerable judicial attention" (Chrysler Motors Corp. v. Auto Body Panels of Ohio, 908 F.2d 951 (1990)).

[132]Chrysler Motors Corp. v. Auto Body Panels of Ohio, 908 F.2d 951 (1990).

[133]The action was brought in front of the International Trade Commission (ITC) under Section 337 of the Tariff Act of 1930, see In re certain Automotive Parts, Investigation Nr. 337 – TA-557, 2007 WL 2021234 (ITS 2007), in the U.S. International Trade Commission. The allegedly infringed patents were two designs of its front bumper fascia, two versions of its rear bumper fascia, its hood, its fender, its side-view mirror and its tail lamp.

[134]Notice of Final Determination of Violation of Section 337 and Issuance of General Exclusion Order (https://www.usitc.gov/publications/337/pub4012.pdf).

enforceable and infringed.[135] Based on this ruling, the ITC issued an Exclusion Order[136] prohibiting entry of the infringing parts into the U.S. Ford filed a second similar action with the ITC in 2008. Both disputes were settled by allowing LKQ to distribute Ford spare parts in the U.S.[137]

In the past few years Ford, Chrysler and others have filed several infringement. Many are still pending at the time of this writing and so far none has resulted in a decision on the merits.[138]

In a new twist to the spare parts disputes, the Automotive Body Parts Association (ABPA) filed an action against Ford Global Technologies in 2014, challenging the validity of several of Ford's design patents on collision repair parts, including hoods, mirrors, grills, head lamps and fenders, on the grounds of exhaustion and functionality.[139] The fact that this action, filed on behalf of and representing the interests of the spare parts manufacturers, denotes both the magnitude of the issue and the economic strength and sophistication of the independent

---

[135]Patents held invalid: – '1 19, '912, and '979; patents held valid and infringed: '890, '552, '135, '579, '561, '044, and '615. Decision at In re Certain Automotive Parts, 2007 WL 2021234 (ITC 2007).

[136]6 June 2007, available at: https://www.usitc.gov/publications/337/pub4012.pdf.

[137]A settlement was entered under which Ford granted defendant LKQ permission to distribute parts protected by Ford design patents (Hawker 2010, pp. 6 et seq.).

[138]Ford Global Technologies, LLC v. New World International, et al., filed 29 January 2015 (E.D. Texas, see https://search.rpxcorp.com/lit/miedce-298368-ford-global-technologies-v-new-world-international); Ford Global Technologies, LLC v. New World International, Inc., Auto Lighthouse Plus, LLC, and United Commerce Centers, Inc., Case Number 2:15cv10394 (E.D. Mich), filed 14 April 2015 (see https://search.rpxcorp.com/lit/txndce-258581-new-world-international-v-ford-global-technologies 2:15-cv-10394); Chrysler Group LLC v. LKQ Corporation et al., 2:14-cv-10210, E.D. Michigan, filed 16 January 2014; January 2014 – action by Chrysler against LKQ/Keystone for infringement of ten design patents relating to various LKQ vehicle parts, including tail lamps, headlamps, a mirror, a grille, a bumpers, and fenders of the Ram truck. The action settled in June 2014, by Chrysler licensing the design patents to LKQ for distribution in the U.S.; CFM Consol Inc. v. Specialty Auto Parts USA, Inc., Nr. 3:12-cv-05923 (W.D. Wa.) (12 October 2012), infringement action for its design patent directed to a combined fan assembly and shroud; Toyo Tire & Rubber Co., Ltd. v. Atturo Tire Corp., Nr. 1:14-cv-00206 (N.D. Ill. 13 January 2014) asserting design patent infringement; Toyo Tire and Rubber Co., Ltd. v. S. China Tire and Rubber Co., Ltd. l, 8-14-cv-00024 (C.D. Ca. 7 January 2014) asserting design patent infringement.

[139]Automotive Body Parts Association v. Ford Global Technologies, filed in the ED Tex, 2013, seeks declaratory judgment of invalidity of certain design patent for certain body parts, D489299 (Exterior of Vehicle Hood); D489657 (Vehicle Side View Mirror); D494517 (Vehicle Side View Mirror); D496890 (Vehicle Grill); D501685 (Vehicle Head Lamp); and D508223 (Fender). Some of these patents were dropped and upon transfer to DC Michigan only two patents remained at issue D489299 and D501685 (Vehicle Head Lamp, Exterior of Vehicle Hood); order to transfer (http://law.justia.com/cases/federal/district-courts/michigan/miedce/2:2015cv10137/297899/3/). See Automotive Body Parts Association v. Ford Global Technologies, LLC, Dkt#48 (E.D. Tex., filed October 3, 2014). Memorandum and Order, 7 January 2015, granting transfer to E.D. Michigan (http://law.justia.com/cases/federal/district-courts/michigan/miedce/2:2015cv10137/297899/3/).

spare parts industry. It may signal a shift to the courts of the so far unsuccessful efforts towards a legislated repair clause.

Finally, taking the debate to yet a new level, Daimler AG filed lawsuits against Amazon and several other online retailers.[140] The complaint alleges infringement of about 20 of Daimler's design patents (as well as of several trademarks) by wheel replicas sold by the named respondents in the U.S.[141] The actions are filed in different venues, including the Western District of Washington and the International Trade Commission and seek a general exclusion order for the infringing products as well as damages.[142] In June 2016 the ITC voted to launch an investigation.[143]

Overall, litigation surrounding the spare parts issue in the U.S. presents a fundamentally different picture from that in the EU. With no statutory repair clause, fewer actions pending and virtually no decisions on the merits, the field remains wide open. The number and nature of recently filed lawsuits, however, indicates that changes can be expected.

5.3.3 South Africa—A Judicially Created Repair Clause?

Yet another approach to the spare parts dispute comes from South Africa. South Africa's design law protects aesthetic designs under the South Africa Designs Act of 1993. Spare parts are protectable as aesthetic designs, if they are new and can be judged solely by the eye.[144] Spare parts per se are deemed functional and therefore

[140]ITC Investigation Regarding Certain Passenger Vehicle Automotive Wheels, number 337-TA-1006, instituted upon a complaint by Daimler AG; Daimler AG v. Amazon.com Inc., case number 2:16-cv-00518, W.D. Wash, filed April 2016. The complaint names as defendants: A-Z Wheels, d/b/a UsaRim/UsaRim.com; Eurotech Wheels; Galaxy Wheels and Tires, LLC; Infobahn International, Inc. d/b/a Infobahn/Eurotech/Eurotech Luxury Wheels/Eurotech Wheels/Usa Rim; Amazon.com Inc; A SPEC Weheels & Tires LLC d/b/a A SPEC Wheels & Tires; American Tire Distributors Holdings, Inc.; American Tire Distributors, Inc; Onyx Enterprises Int'l Corp d/b/a CARID.COM; O.E. Wheel Distributors, LLC; Powerwheels Pro, LLC and Trade Union International inc. d/b/a Topline. See http://www.law360.com/articles/807711/itc-launches-probe-of-possibly-counterfeit-daimler-wheels.

[141]Daimler AG v. Amazon.com Inc., Case number 2:16-cv-00518, W.D. Wash, filed April 2016 (see http://www.geekwire.com/2016/daimler-ag-sues-amazon-knockoff-mercedes-wheels-new-test-e-commerce-giants-liability-counterfeit-goods/).

[142]The outcome of these actions is yet to be determined. The ITC complaint is available at: http://www.itcblog.com/images/daimlercomplaint.pdf. The ITC actions assert the intent to file similar actions against these respondents in eight different District Courts in the U.S.

[143]See http://www.tirebusiness.com/article/20160616/NEWS/160619951/itc-begins-investigation-of-passenger-wheels.

[144]"(A)ny design applied to any article, whether for the pattern or the shape or the configuration or the ornamentation thereof, or for any two or more of these purposes, and by whatever means is applied, having features which appeal to and are judged solely by the eye, irrespective of the aesthetic quality thereof".

not registrable,[145] unless they contain aesthetic features. In that case they may be registered as aesthetic designs, as it is recognized that products may have both aesthetic and functional features. South Africa has not adopted an exclusion for visible spare parts for purposes of repair to restore a product's original appearance, comparable to Article 110(1) CDR.

Against this background, BMW filed an action against Grandmark for infringement of various design-protected automotive parts, including a bonnet, a grille, a headlight assembly, and a front fender. Grandmark responded by seeking the revocation of BMW's design protection, on the grounds of functionality and lack of aesthetic features. The court of first instance held BMW's designs invalid and unregistrable as aesthetic designs.[146]

This ruling was affirmed by the South African Supreme Court, based on the fact that under South African design law, aesthetic designs are those that invite customer selection and discrimination based on their visual appeal. Their protection is premised on a consumer's selection of the particular design as a result of its appeal to the eye. While such appeal may exist with respect to an entire car, it does not exist with respect to component parts. Component parts are selected by customers not for their appeal to the eye, but "*solely* for the function they perform – which is to replace a component so as to restore the vehicle to its original form".[147] Customers often do not see the component before it is fitted to the vehicle. Consequently, the design of these parts is purely functional and cannot be registrable as aesthetic design.[148]

This narrow view of functionality seems surprising even under South African law. A broader view might have found that the parts at issue have some aesthetic features, after all they are part of the same design that prompts the customer's selection of the vehicle in the first place. Acceptance of this fact might have rendered the parts registrable, particularly since South African law recognizes the presence of both aesthetic and functional features in one design.

The overall impact of the Grandmark decision remains subject to some debate.[149] One view holds that most component parts are incapable of being protected by design. This is so because they merely serve the function of replacing other parts and are not selected by the customer for their aesthetic value. Under this

---

[145]Section 14(6) South Africa Designs Act 1993 (Act Nr. 195 of 1993), as most recently amended 1 May 2011.

[146]BMW v. Grandmark International (Pty) Ltd and Another (722/12) [2013] ZASCA 114; 2014 (1) SA 323 (SCA), 18 September 2013, Supreme Court of Appeal, available at: http://www.saflii.org/za/cases/ZASCA/2013/114.html.

[147]BMW v. Grandmark International (Pty) Ltd and Another (722/12) [2013] ZASCA 114; 2014 (1) SA 323 (SCA), 18 September 2013, Supreme Court of Appeal.

[148]BMW v. Grandmark International (Pty) Ltd and Another (722/12) [2013] ZASCA 114; 2014 (1) SA 323 (SCA), 18 September 2013, Supreme Court of Appeal: If parts were interchangeable, such as mirrors, wheels, seats, steering wheels, they might be eligible for protection.

[149]Cf. Magagula 2012.

view, the South African Supreme Court has judicially created a repair clause and opened up the spare parts market to competition.[150]

Others approach the matter more cautiously, limiting the effect of the decision to its facts. It must be kept in mind that South African law does not provide for substantive examination of design applications. The court might be therefore merely testing the validity of the individual registration at issue. Because BMW was unable to provide the requisite evidence of novelty and aesthetic features, the registrations were invalidated.[151]

5.4 Brazil[152]

Brazil protects industrial designs which are ornamental, new and original, i.e. the design must display a distinctive visual configuration.[153] The rights are granted automatically, upon application without substantive examination. Brazil has not adopted a provision in the nature of a repair clause.

However, design rights in spare parts have been challenged in front of the Brazilian competition authority, the Administrative Council for Economic Defense (Conselho Administrativo de Defesa Econômica, CADE). Thus, for instance, an action brought in 2006 against Alcoa, resulted in a finding that by asserting its IP right, Alcoa unlawfully exercised it. The decision that was subsequently affirmed by CADE's full council.

More recently, such actions have been unsuccessful.[154] In 2007, Brazil's National Association of Independent Car Parts Manufacturers (Associaçao Nacional dos Fabricantes de Autopeças, ANFAPE) complained to the competition council's Economic Law Secretary (Secretaría de Economía, SDE) about the conduct of Volkswagen, Ford and Fiat, claiming that their assertion of intellectual property rights constituted abuse of economic power. The SDE ruled in favor of the auto manufacturers, on the ground that intellectual property rights are not limited to the primary market. The two markets must be viewed as interdependent and exercise of these rights in the aftermarket was justified, in order to enable auto manufacturers to recoup the R&D investments made in the primary market.[155] An appeal followed. The CADE Attorney General affirmed, holding that the Brazilian Industrial Property Act does not exempt from protection spare parts used for

[150]For details, see Magagula 2012.

[151]Magagula 2012.

[152]The section on the legal situation in Brazil is based on the contribution of Professor Fabiano Teodoro de Rezende Lara, Universidade Federal de Minas Gerais, Brazil.

[153]Articles 96 et seq. Brazilian Industrial Property Law (nº 9.279/96).

[154]CADE, Fragoso v. Alcoa (2006).

[155]Averiguação Preliminar Nº do Processo: 08012.002673/2007-51; Associação Nacional dos Fabricantes. de Autopeças – ANFAPE x Fiat Automóveis S.A., Ford Motor Company Brasil Ltda., Volkswagen do Brasil Industria de Veículos Automotores Ltda.

purposes of restoration of the original appearance of a complex product and that the auto manufacturers legitimately relied on their IP rights.[156] Thereafter the CADE Board ordered initiation of an administrative proceeding based on alleged abusive exercise of IP rights this writing.

In the meantime, industrial design protection for spare parts remains in effect in Brazil. Auto manufacturers are free to register design rights and assert them against aftermarket producers. While some Brazilian judges have ruled against the independent car parts manufacturers, issuing injunctions against spare parts manufacturers,[157] at least one challenge to the validity of industrial design protection for spare parts has been rejected.[158]

5.5 Japan

Designs are protected under the Japan Design Act of 1959 if they are embodied in a movable object, an "article" in the wording of the relevant Article 2(1),[159] and if this article "produces an aesthetic impression on the sense of sight".[160] Since a revision of the Design Act in 1999, the term "article" explicitly includes a "part of an article".[161] Japanese design law does not contain an explicit provision on complex products. Instead, they are treated as any other "article" in the sense of Article 2(1). With regards to an older judgement of Tokyo High Court, it can be

[156] Averiguação Preliminar Nº do Processo: 08012.002673/2007-51; Associação Nacional dos Fabricantes. de Autopeças – ANFAPE x Fiat Automóveis S.A., Ford Motor Company Brasil Ltda., Volkswagen do Brasil Industria de Veículos Automotores Ltda.

[157] The Brazilian Federal Court of the 2nd Region rejected the claim of nullity of the industrial designs made by Orgus against Ford Motor Company, reinstating the validity of the registrations (Tribunal Regional Federal da 2ª Região, 1 August 2013, Apelação nº 080932630.2010.4.02.5101). São Paulo State Court enjoined Universal Industria Metalurgica Ltda., an independent manufacturer of door handles, from manufacturing the handles that infringed industrial design rights held by Ford Motor Company Brasil Ltda., Tribunal de Justiça do Estado de São Paulo, 9 August 2011, Apelação nº 9184726-44.2009.8.26.0000, da Comarca de São Paulo.

[158] Autimex Comércio e Importação Ltda. and Plastiron Industria e Comercio Ltda. were enjoined from producing bumpers with industrial design rights held by Ford Motor Company Brasil Ltda., Tribunal de Justiça do Estado de São Paulo, 4 October 2011, Apelação nº 0302734-36.2009.8.26.0000, da Comarca de São Paulo.

[159] Article 2(1) Japan Design Act (Act Nr. 125 of 13 April 1959): "design" shall mean the "shape, patterns or colours, or any combination thereof, of an article (*including a part of an article* (…)), which creates an aesthetic impression through the eye" (emphasis added).

[160] For details, see Hallenborg et al. 2008, pp. 99 et seq.; see also WIPO 2007, p. 224.

[161] See Hallenborg et al. 2008, p. 99.

assumed that must match parts fall under the term "article" as well.[162] Consequently, Japan is one of the countries that do not except spare parts from protection. Whether this fact gives rise to a public debate, that could possibly lead to a change in the law, remains uncertain.[163]

5.6 Interim Conclusion

Review of the cases in various jurisdictions may at first glance create the impression of a disorganized patchwork. Closer examination, however, points to the fact that common economic realities and business interests across jurisdictions, largely related to the overarching goal of capturing as much of the respective spare part market as possible. These result in similar underlying tensions among stakeholders and in predictable enforcement patterns. Lawsuits are brought in most cases by OEMs seeking to enforce design rights in spare parts. Invariably they are opposed by independent spare part manufacturers, either by asserting a repair clause defense, or by challenging the validity of the OEM's IP rights. In the U.S., challenges to validity appear to be more frequent. More recently, in the U.S., independent spare parts manufacturers have used invalidation attempts offensively, by filing declaratory relief actions to invalidate design rights. In Brazil, independent spare parts manufacturers are pursuing challenges to IP rights in front of competition authorities, actions which are opposed by OEM's relying, inter alia, on the strength of their IP rights.[164]

In terms of substantive law, several conclusions have emerged. In regimes in which a statutory repair clause is in effect,[165] some of the most important points are: (1) there is no record of a successful challenge to the validity of the repair clause, (2) the repair clause is to be construed narrowly, (3) Article 110(1) CDR is assertable as a defense against community design rights, but not against rights under national design law,[166] (4) the repair clause cannot be asserted as a defense against other IP rights, specifically trademark rights (and most likely other IP rights as well), (5) the repair clause applies to parts intended to restore a complex product's original appearance, and not to parts that serve to upgrade the appearance of a product, based on the reasoning that if a spare part's appearance is not dependent

---

[162]See Kur and György 2016, pp. 23 et seq.

[163]For more details, see Kur and György 2016, pp. 23 et seq.

[164]Averiguação Preliminar Nº do Processo: 08012.002673/2007-51; Associação Nacional dos Fabricantes. de Autopeças – ANFAPE x Fiat Automóveis S.A., Ford Motor Company Brasil Ltda., Volkswagen do Brasil Industria de Veículos Automotores Ltda.

[165]These decisions hail largely from the EU (Germany, Italy, Spain, UK), providing an interpretation of Article 110(1) CDR, or national clauses modelled on it.

[166]LG Düsseldorf, 28 November 2013, 14 c O 304/12 U, GRUR-RS 2014, 00345; see also headnote, as formulated by Ebert-Weidenfeller 2014, p. 56.

on the complex product's appearance (or there is no objective necessity for it to conform to that appearance), the consumer has a choice of parts and the producer of such parts does not monopolize the market. This rule has been applied to wheel caps and wheel rims, but conceivably its logic can extend to other parts as well, and may serve to progressively reduce the scope of the repair clause. Australia has adopted a repair clause, but no decisions have been reported.

In regimes in which no statutory repair clause is in effect,[167] lawsuits by OEMs asserting their design rights have been challenged in various ways. The primary defense asserted by spare parts manufacturers relates to validity of the spare part's design.[168] In the U.S. and South Africa, a counterclaim for invalidity is asserted in virtually every case. Enforcement actions are therefore more risky for OEMs, as an adverse ruling on validity would mean loss of the IP right in that jurisdiction. To what extent a ruling of invalidity transfers to other jurisdictions is uncertain. As relates to anticipation, a previously unknown piece of prior art, signaled by a decision in one jurisdiction, would likely be considered in another jurisdiction as well. As relates to functionality, the substantially different standards under e.g. U.S., EU or South African design laws would probably prevent decisions from other jurisdictions even from serving as guidance.

References

Beier FK (1994) Der Musterschutz von Ersatzteilen in den Vorschlägen für ein Europäisches Musterrecht. GRUR Int 716–732

Beyerlein T (2016a) Aktuelle Entwicklungen im Designrecht. Mitt 107:306–310

Beyerlein T (2016b) Die Reparaturklausel des Designrechts als (vermeintliche) Schranke des Markenrechts. MarkenR 18:196–199

Brtka R, Soetens, R (2015) Focus on the automotive industry: the protection of spare parts using Community designs In: Lexology. http://www.twobirds.com/en/news/articles/2015/global/designwrites-mar/focus-on-the-automotive-industry-the-protection-of-spare-parts-using-community-designs

Drexl J, Hilty R, Kur A (2005) Design protection for spare parts and the Commission's proposal for a repairs clause. IIC 36:448–457

Ebert-Weidenfeller A (2014) Reparaturklausel des Art. 110 GGV bei Verletzung deutscher Geschmacksmusterrechte nicht anwendbar. GRUR-Prax 56

Eichmann H (1996) Das europäische Geschmacksmusterrecht auf Abwegen? GRUR Int 859–876

Eichmann H (1997) Kein Geschmacksmusterschutz für must-match-Teile? GRUR Int 595–609

Eichmann H, Vogel von Falckenstein R, Kühne M (eds) (2015) Geschmacksmustergesetz. Kommentar, 5th edn. C.H. Beck, München

[167] Judicial activity in such jurisdictions hails primarily from the U.S., with a single case from South Africa.

[168] It is less likely to be raised against the introduction of a repair clause into EU law, as the issue would necessarily not be before the court in an infringement claim, and may be the subject of a separate proceeding.

Ferretti N (2016) Wheel rims and repair clause, two referrals to the cjeu: which orientation will prevail? ECTA Bull October:22

Fezer KH (2016) Theorie der Funktionalität der Immaterialgüterrechte als geistiges Eigentum. Zugleich eine rechtstheoretische Grundlegung zum Vorabentscheidungsverfahren „Ford/ Wheeltrims“. GRUR 30–38

Fitzpatrick JF (1989) industrial design protection and competition in automobile replacement parts-back to monopoly profits? U Balt L Rev 19:233–270

Grabrucker M (2015) Stellungnahme zum Vorlageersuchen an den Europäischen Gerichtshof (C-500/14) des italienischen Tribunale di Torino sezione specializata in material di imprese (Anche quale Tribunale dei marchi comunitari) betreffend ein Verfahren über die Verletzung von Markenrechten der Fa Ford Motor Company gegen Wheeltrims s.r.l. (Ersatzteilehersteller) (To be published)

Guizzardi S (2005) Reparaturklauseln und Eintragungsfähigkeit von Ersatzteilen in Italien – Probleme und Perspektiven. GRUR Int 299–304

Hackbarth R (2015) Reparaturklausel im Markenrecht nicht anwendbar. GRUR-Prax 481

Hackbarth R (2016) EuGH-Vorlage zur Reichweite des Art. 110 GGV. GRUR-Prax 382

Hallenborg L, Ceccagnoli M, Clendenin M (2008) Chapter 3 Intellectual property protection in the global economy. In: Libecap GD, Thursby M (eds) Advances in the study of entrepreneurship, innovation, and economic growth. Emerald, Bingley (UK)

Hartwig H (2016) Spare parts under European design and trade mark law. GRUR Int 102–108

Hawker N (2010) The automobile aftermarket: crash parts, design patents, and the escape from competition. http://www.antitrustinstitute.org/sites/default/files/aai%20collision%20repair%20parts%20commentary_032220101350.pdf

Janis MD (1999) A Tale of the Apocryphycal Axe: Repair, Reconstruction, and the Implied License in Intellectual Property Law. Maryland L R 58:423–527

Klawitter C (2015) „Italienischer Torpedo“ und Reparaturklausel in Article 110 GGV.GRUR-Prax 466

Kroher J (1993) EG-Geschmacksmusterschutz für Kraftfahrzeug-Ersatzteile. GRUR Int 457–465

Kur A (2016) Ersatzteilfreiheit zwischen Marken- und Designrecht. GRUR 20–30

Kur A, György A (2016) Protection of Spare Parts in Design Law—A Comparative Law Analysis (To be published)

Larenz K, Canaris CW (1995) Methodenlehre der Rechtswissenschaft, 3rd edn. Springer, Berlin, Heidelberg

Magagula L (2012) Are brand-owners caught in a pickle? BMW v. Grandmark, Norton Rose Fulbright. http://www.nortonrosefulbright.com/knowledge/publications/71305/are-brand-owners-caught-in-a-pickle-bmw-v-grandmark

Osborne JW (2004) A coherent view of patent exhaustion: a standard based on patentable distinctiveness. Santa Clara High Technol L J 20:643–693

Pentheroudakis C (2002) Die Umsetzung der Richtlinie 98/71/EG über den rechtlichen Schutz von Mustern und Modellen in den EU-Mitgliedstaaten. GRUR Int 668–682

Ruhl O (2010) Gemeinschaftsgeschmacksmuster. Kommentar, 2nd edn. Carl Heymanns Verlag, Köln

Saidman PJ (2009) Functionality and Design Patent Validity and Infringement. J Pat & Trademark Off Soc'y 91:313

Salonen S (2016) Rims hit regulation: replica rims and the spare part exception. http://www.castren.fi/blogandnews/blog-2016/rims-hit-regulation-replica-rims-and-the-spare-part-exception/

Stoate N, Smith C (2005) A nice little earner. The after sales market in the automotive industry. Copyright World 155:16–18

Straus J (2005) Ende des Geschmacksmusterschutzes für Ersatzteile in Europa?. Das Mandat der Kommission und seine zweifelhafte Ausführung. GRUR Int, Vorgeschlagene Änderungen der EU-Richtlinie, pp 965–979

Wilbanks KB (2013) The challenges of 3D printing to the repair-reconstruction doctrine in patent law. George Mason L R 20:1147–1181

Würtenberger G, Loschelder M (2015) Stellungnahme des GRUR-Fachausschusses für Designrecht unter Mitwirkung von Mitgliedern des Fachausschusses für Wettbewerbs- und Markenrecht zum Vorabentscheidungsersuchen des Tribunale Ordinario di Torino „Ford Motor Company/Wheeltrims SRL" (Rs. C-500/14). GRUR 348–351

Chapter 6
Considerations Towards a Compromise

6.1 Introduction: Taking Stock

The picture that emerges from the above review is, on the one hand, that of a political impasse in legislatures, and on the other, mounting tension among the stakeholders, resulting in an increasing number of lawsuits.

The European Commission withdrew its proposal to amend the Directive on 21 May 2014, following the veto by France, Germany, Slovakia, Romania, the Czech Republic and Sweden in the European Council in 2008.[1] No change of position is expected in the key votes of large automotive manufacturing states, such as Germany and France. Germany's informal arrangement with the auto industry for non-assertion of design rights does not require it to change the *status quo*. France indicated its position on 29 January 2014, by voting down the repair clause in its national legislation,[2] even though adoption of the clause had been recommended by the French Competition authority on 8 October 2012.[3] Other countries with developed automotive industries, such as Romania, have not adopted the clause, despite recommendations from their Competition Authorities.[4] In short, the result of a future Council vote would likely be the same and an agreement on the issue in the near future is improbable.[5] In the U.S., the situation fares no better. Legislative proposals for introduction of a repair clause have been made year after year without being adopted.

Nonetheless strong tensions among stakeholders underlie this apparently static situation. They occasionally "bubble up" in the form of legal actions with a noticeably accelerating trend. In the EU, national laws of member states remain non-harmonized. While Article 110(1) CDR applies in relation to Community

[1]Meeting the requirements for a blocking minority under Article 16(4) TEU, Article 238(2) TFEU.
[2]See Loi n° 2014-344 du 17 mars 2014 relative à la consummation (Loi Hamon).
[3]Autorité de la concurrence 2012.
[4]See Romanian Competition Authority Recommendation, 2012.
[5]Drexl 2012, p. 370.

D. Beldiman and C. Blanke-Roeser, *An International Perspective on Design Protection of Visible Spare Parts*, SpringerBriefs in Law,
DOI 10.1007/978-3-319-54060-3_6

design rights, it does nothing to exempt parts from protection at national level in countries which have not separately adopted the repair clause.[6] The number of lawsuits relating to spare parts filed throughout European countries is on the rise.[7] In the U.S., the number of lawsuits relating to design protection of spare parts has also been steadily increasing.

Against this background, an outright adoption or rejection of the repair clause as formulated in the various proposals made is unlikely. However, if the negotiations were expanded to include further issues of concern to the stakeholders and those concerns could be addressed as a package in conjunction with the repair clause, the chances of a compromise solution might increase. Without seeking to be exhaustive, the following will attempt to lay out a few of the considerations which might be included in a possible package.

6.2 Limited Term of Protection Within a Repair Clause

A variation in the term of the exemption could enhance the overall acceptability of a repair clause. The exemption might, for instance, commence three years after registration of the design. This would give the OEM a period of exclusivity, while also opening up the market to competition upon expiration of the initial protection period.

The precise duration of a possible protection period hinges on the stakeholders' ability to exploit spare parts in the secondary market. It must achieve a meaningful allocation of benefits among stakeholders. A multitude of complex factors comes into play in determining the optimal duration. One such factor is whether the repair clause should be limited to motor vehicles or if it should extend to other products. Arguably, in the high volume motor vehicle market rightholders are able to recoup their investment sooner than in a market for products with a low sales volume. If the repair clause were limited to the automotive industry, this fact might suggest a

[6]This clause has been adopted as a transitional provision (see Article 110(1) CDR). There has been no movement towards amending the Design Directive lastly; therefore, from a legislative perspective, the freeze-plus solution remains in effect. This leaves Article 110(1) CDR as the currently available model for a repair clause and the interpretation of the article proposed by the European Commission, which will therefore constitute the focus of the following discussion. According to Drexl 2012, p. 372, and Zwanzger 2010, p. 3, one of the reasons for declining to adopt the clause was uncertainty as to whether the repair clause has the ability to restrict competition in industries other than the automotive, such as kitchen appliances, watches and furniture. The study commissioned by the EU on the Economic Effects of Industrial Design (Europe Economics 2015) gave some consideration to the watch industry, but did not seem to yield conclusive results and will therefore not affect the vote on this matter.

[7]The Judgement Helsinki District Court, 19 November 2015, Nr. 15/149362, R 14/5257 – Public Prosecutor and BMW v. Hotari and Pohjola mentions 40 decisions on wheel caps alone.

longer exemption period. Another factor is the product's lifecycle.[8] At a certain vehicle age, natural wear and tear increases and, with it, the need for spare parts. Typically, "wear" parts, need to be replaced after a given period of time and the need to do so varies depending on the time in the vehicle's life, presumably increasing over time.[9] This would suggest that each stakeholder would wish to vary the timing of the exemption in a manner that would allow them to exploit the spare parts market when the need for such parts is greatest. This however, does not take into consideration purely "crash" parts, where the need is unrelated to the product's age and may be required from the first day a new car model is on the road.[10] A further factor is the duration of the warranty. During the warranty period spare parts are largely supplied by the OEMs. In theory therefore, independent suppliers would not be substantially disadvantaged by the delay in the start of the exemption period. However, independent suppliers argue that at the end of the warranty period they might have difficulties winning back the customer.[11]

In short the question of a shortened exemption and its onset remains a controversial one and not a prospect of easy compromise.

6.2.1 EU Legislative Efforts and the Stakeholders' Positions

The option of a shortened exemption was first considered by the EU in 1993,[12] in an effort to reach an agreed-upon version for purposes of Directive. The proposal was for a three-year protection term followed by an exemption. This model splits exploitation of the aftermarket between OEMs and spare parts manufacturers; it makes inroads into the OEM's IP right, on the one hand, and avoids consumers being tied to a particular manufacturer indefinitely, on the other.[13] Ultimately, however, this approach was not adopted. The EU Impact Assessment cautioned that in the longer term, the "left-over" part of the market might not be sufficient to

[8]See ECTA, Design Committee, Position Paper Designs and Spare Parts, 2016; European Commission, Proposal for a Directive amending Directive 98/71/EC 2004. In the field of motor vehicles, numbers between 7 and 13 years are being mentioned. Data on this are unclear: the ECTA proposal is based on an assumption of 7 years, other sources cite 13.

[9]"(B)ody work segment represents 55% of vehicle repairs and maintenance turnover for vehicles aged under two years and only 15% of vehicle repairs and maintenance turnover for vehicles aged over 10 years" (Autorité de la concurrence 2012, pp. 237 et seq.).

[10]European Commission, Proposal for a Directive amending Directive 98/71/EC 2004.

[11]European Commission, Proposal for a Directive amending Directive 98/71/EC 2004.

[12]Proposal for a European Parliament and Council Directive on the legal protection of designs of 3 December 1993, supra section 4.2.1.1.

[13]See British Leyland Motor Corp. v. Armstrong Patents Co. [1986] UKHL 7 (27 February 1986); BMW AG v. Round and Metal Ltd and Philip David Gross [2012] EWHC 2099 (pat).

incentivize new entry.[14] Compared to "full liberalization (…) this option would be likely to result in higher prices and lower employment".[15]

Stakeholders remain split on the issue. Their arguments largely parallel the arguments made in support and against adoption of the clause itself. OEMs take the position that an additional period of exclusivity would be justified to obtain the design premium to which they are entitled, on the assumption that the investment has not been fully amortized in the primary market.[16] It would help avoid adverse consequences such as diminished incentive due to reduced profit and consequent unwillingness to further invest into product innovation. Some of the countries which have not adopted the clause perceive the limited duration option as preferable to their current arrangements, as, at least in the longer term, it raises the prospect of competition.[17] Spare parts providers, on the other hand, point to the adverse effects that even such a limited protection period would have. A shortening of the exemption period would reduce the spare parts manufacturers' profits on their investment. In addition, independent service providers would have trouble converting consumers to the use of independent facilities, after a period of exclusivity during which repair needs are met by the authorized facility.[18] All of this would make the business of spare parts unattractive and, indeed, might discourage some from entering the market altogether.[19] Effectively, it is argued, even a shorter period of exclusivity, would de facto result in a "full protection" scenario. Finally, from the public's perspective, protection, even a limited period, would still result in increased prices.[20]

6.2.2 *Models of Partial Exemption Periods*

Reduced terms of protection adopted or suggested under this model have ranged between 2.5 and 5 years. Greece has, for instance, adopted the repair clause proposed in an early version of the Directive proposal, which limited protection to five years from the launch of the product. Upon expiration of the five years, the market

---

[14]European Commission, Proposal for a Directive amending Directive 98/71/EC 2004.

[15]Europe Economics 2015, p. 140.

[16]See Emission Impossible, The Economist, 15 March 2016, available at: http://www.economist.com/news/business/21693947-german-carmaker-will-escape-its-emissions-scandal-largely-unscathed-bad-news: "(I)n 2014 the brand's (VW's) profits essentially came from parts sales and royalties". But cf. Drexl et al. 2005a, b, pp. 449 et seq., who proceed on the contrary assumption. However, this assumption is not substantiated by data. A limited term would allow the OEMs to recoup their investment in the "intellectual effort" associated with the design (cf. European Commission, Proposal for a Directive amending Directive 98/71/EC 2004).

[17]Europe Economics 2015, p. 140.

[18]European Commission, Proposal for a Directive amending Directive 98/71/EC 2004.

[19]Autorité de la concurrence 2012, p. 237.

[20]Autorité de la concurrence 2012, p. 237.

was to be open to competition.[21] In addition, the Greek law also required "fair and reasonable remuneration" of the rightholder. Possibly, because of these multiple requirements no conclusive assessment of the duration has been reached.

Interesting in this regard is the evolution of the legislative proposals in the U.S. Initial versions of the repair clause submitted to the U.S. Congress in the 2008/2009 congressional session fully exempted spare parts from design protection from the date they were placed on the market.[22] Subsequent versions became narrower, granting a 2.5 year protection period, followed by an exemption for the remainder of the protection period. Presumably, this protection period was introduced in an effort to reach a compromise. Notably, more recent versions, including the one presented in the 2015/2016 legislative session, further carve up the exemption. Independent providers may "make, offer for sale and import" spare parts at any time after the product has first been offered for sale, whereas use and sale by such providers may only start 30 months thereafter. "Making" has been defined to include testing.[23] The intent in exempting the testing and pre-sale activities from protection appears to be, on the one hand, an effort to incentivize innovation by independent manufacturers and, on the other, to seek a compromise position from OEMs by granting them a period of market exclusivity. So far however, no conclusion has been reached as to these models: the effect of the Greek repair clause is unknown and the various versions proposed in the U.S. have not yet been adopted by Congress. Consequently, the question remains open.

Proposals continue to be made. A 2016 position paper by the European Communities Trademark Association (ECTA) further analyzed the issue and proposed an alternative option of reducing "the term of spare parts design protection to 5 years (with no possibility of extension)".[24] This solution "will allow to compensate the designer's investment and to stimulate its intellectual creation (the true goal of IP) and simultaneously to allow the competitors to enter the market and be compensated for that." On the assumption that the life of a car is seven years, market entry by competitors five years following the product's launch was felt to constitute sufficient time to stimulate competition, while at the same time,

---

[21]Article 26(4) of Presidential Decree No 259/1997 of 19 September 1997: "The rights conferred by the registration of a design or model shall be exercised by third parties who, five years after the first putting on the market of the product to which the design or the model is applied and following the payment of a fair and reasonable remuneration to the owner as agreed by the parties or in case of dispute, as decided by the competent Court provided in the present Presidential Decree, may use it on the following terms: (a) the product in which the design or the model is incorporated is a spare part of a mechanically operated vehicle, and (b) said use is intended to allow the repair of a mechanically operated vehicle, and (c) the public shall be informed of the origin of the product used for the repair under a permanent indication, such as the affixation of a commercial trademark or trade name or by any other suitable means".

[22]H.R. 3059, 111th Cong. (2009); H.R. 5638, 110th Cong. (2008).

[23]"The PARTS Act" H.R. 1057 proposed to amend 35 U.S.C. § 271 in the 2015/2016 congressional session (previously proposed 2012/2013).

[24]See ECTA, Design Committee, Position Paper Designs and Spare Parts, 2016.

rewarding the automotive industry with a monopoly for at least, a limited time frame.[25]

Results somewhat equivalent to an abbreviated exemption period can be achieved by other means as well. For instance, laws might require use of original spare parts during a specified period after the date of the product's launch. In one California state bill proposal, the warranty period was used as measure for duration. The law would have penalized insurers for installing non-OEM parts on vehicles still under warranty.[26]

Whether as part of the repair clause or as a separate measure, the element of duration of protection allows the requisite fine-tuning and thus remains a significant bargaining chip in an attempt to reach a compromise.

6.3 Transition Regime

6.3.1 General Considerations

Gradual implementation of the repair clause might ease its acceptance by some of the stakeholders. A measure of this type has been proposed in France, in the form of a transition period[27] and could serve as a model for other jurisdictions as well.

A number of considerations speak in favor of such a measure. Policymakers are concerned that absent a preparatory period, national spare parts manufacturers and potential market entrants would be unable to gear up their operations in time to compete. Suppliers from countries with a repair clause tradition, might rapidly gain ground in the newly liberalized environment and impact primarily SMEs interested in entering the spare parts market. In a non-liberalized country, a transition period would help the national spare parts industry gear up.[28] A transition period could also benefit vehicle manufacturers.[29] Introduction of a repair clause with immediate effect might find OEMs who rely to a large extent on profits from the aftermarket, unprepared for competitive market conditions and would jeopardize their competitiveness.[30] A transition period would allow OEMs time to adapt their economic

[25]See ECTA, Design Committee, Position Paper Designs and Spare Parts, 2016.

[26]E.g. proposed Senate Bill o3 2008; see also Albright 2009.

[27]See generally, Autorité de la concurrence 2012.

[28]France is concerned that if liberalization were imposed upon the French visible spare parts market without allowing French market stakeholders any time to prepare, this could harm the French equipment suppliers present in France or, at the very least, deprive them of opportunities presented by liberalization at a European level. See Autorité de la concurrence 2012, pp. 234, 250.

[29]Vehicles are sold at a loss or at a very low profit margin (Autorité de la concurrence 2012, p. 247).

[30]Autorité de la concurrence 2012, p. 246.

model and reposition themselves in the new market.[31] Of course, a transition period also delays implementation of liberalization. However, in instances in which the legislature is inclined to adopt a clause, a transition period may constitute a useful tool in seeking a compromise.

6.3.2 *Implementation of a Transition Period—The French Model*

In 2012, the French Competition Authority undertook a detailed review of the manner in which a transition provision could be structured. Its main aspects are as follows:

A transition period might either be put into law or take the shape of an informal agreement by OEMs not to assert their design rights in visible spare parts.[32] This latter, de facto approach is easy to implement and provides a simple way to test the effects of liberalization. It would not constitute a threat of legal action against independent spare parts suppliers who are in the process of gearing up for production and would thus not discourage investment. Independent spare parts suppliers would be placed on parity with their counterparts in liberalized countries.[33] The method's downside would be that foreign suppliers would also be allowed to sell within the country. Finding a way to prevent them by way of an informal agreement, as is possible with domestic enterprises, would be difficult.[34] The policy goal of building up a national industry could thus be weakened. Legal enforceability of this option is more limited. However, it would still be feasible if accompanied by a default provision, under which the repair clause would be enacted in case of breach by the industry.[35]

A legislated transition period could take the form of a simple delay in the effectiveness of the law, in order to give national suppliers time to adjust, or alternatively, of a gradual relaxation of the market. The French Competition Authority suggested criteria for a step by step opening of the market, based on (1) the age of the vehicle, (2) the entities impacted or (3) types of spare parts, each of which is described more fully below.[36]

---

[31]"It seems advisable to take the current difficulties into account and to give the French vehicle manufacturers sufficient time to adapt their economic model, instead of immediately depriving them of a source of profits, which could help them to reposition themselves successfully" (Autorite de la Concurrence p. 234).

[32]Autorité de la concurrence 2012, p. 254. This model has been applied in Germany. See supra section 4.2.1.2.2.

[33]Autorité de la concurrence 2012, p. 255.

[34]See for instance the agreement reached between the German government and the German auto industry regarding non-assertion of design rights.

[35]See Autorité de la concurrence 2012, pp. 256 et seq.

[36]Autorité de la concurrence 2012, p. 258.

If based on vehicle age, competition for older vehicles could be allowed first and gradually descend in age. This solution was however perceived as too narrow to sufficiently encourage investment into an independent industry. Furthermore, administrative implementation would be difficult, as would be keeping track of part numbers and vehicle models.[37] Preferable over the first option, a transition limited to original equipment suppliers would engage the entire sector of independent equipment suppliers in preparing for the open market.[38] Even so, because traditionally, independent equipment suppliers are reluctant to compete head-on with OEMs, the measure might not have the desired impact. Finally, liberalization might be phased in, based on types of parts. Thus, for instance, in year one, windows and glass would be liberalized, in year two, mirrors and bumpers would be added and in the following years, sheet metal parts, etc., until the market is fully opened up. If clear deadlines are specified and observed, this solution would offer a substantial level of legal certainty. Both original and independent equipment suppliers would have a greater incentive to invest and compete, resulting in a favorable impact on prices.[39] Job losses with OEMs would, if at all, occur gradually, and would be compensated by the creation of new jobs for competitors.[40] The French Competition Authority viewed this last option as being most beneficial for consumers and the long-term vitality of the industry itself.[41]

In terms of duration, a transition period of four to five years was considered sufficient to encourage the various stakeholders to make changes in anticipation of the opening-up of the market.[42]

To conclude, a transition clause is appealing in terms of allowing the national spare parts industry to gear up. From an OEM's standpoint, a repair clause subject to a transition provision is preferable to a clause with immediate effect. In instances in which the legislature is poised to adopt a clause, a transition provision would therefore render the clause more acceptable to OEMs.

6.4 Disclosure Laws

Disclosure of information about the origin and qualities of spare parts could serve as an important complementary measure to a repair clause, whether as an integral part of it or separate, and facilitate adoption by some of the stakeholders.

[37] Autorité de la concurrence 2012, p. 259.

[38] Autorité de la concurrence 2012, p. 260.

[39] Autorité de la concurrence 2012, pp. 264, 266.

[40] Autorité de la concurrence 2012, p. 266.

[41] See Autorité de la concurrence 2012, p. 266.

[42] This time frame was recommended by the French Competition Authority in its study, see Autorité de la concurrence 2012, p. 252.

6.4.1 *The Need for Disclosure*

A collision spare part may originate from any of the following sources: (1) the OEM, (2) its authorized supplier, (3) an independent supplier supplied by an authorized OEM supplier, possibly from the same production line as the authorized parts (most often these are based on the manufacturer's drawings, specifications and know-how), (4) an unauthorized supplier who reverse engineered the part and attempted to approximate the specifications and general quality as closely as possible, (5) an unauthorized supplier who produces a reverse engineered part solely based on cost considerations, (6) a decommissioned vehicle due to breakage or accident and (7) a salvage yard.[43] In large part, this multitude of possible sources results from the maze of manufacturing and contractual relationships between OEM manufacturers and distributors, and the distributor's ability to supply independent markets.[44]

The origin of a part is likely to affect its quality and constitutes the most relevant factor to consumers next to price. It is therefore important for a consumer to determine whether a part is produced or authorized by the OEM. This was recognized in early drafts of the European Commission's Directive proposal, which included a disclosure of origin requirement.[45] This provision was abandoned in the final version. More recent legislation, however, does not promote clear differentiation. The terms "original", "OEM" and "aftermarket" are used quite loosely. As one U.S. court observed, having found no clarity in dictionary definitions, when it comes to spare parts a latent ambiguity exists between the terms "OEM" and "aftermarket".[46] European law, for instance, defines "original spare parts" as parts "manufactured according to the specifications and production standards" of the vehicle manufacturer.[47] Separately, "spare parts of matching quality" are merely

---

[43]See Testimony of Don Risley, ASA, at Congressional hearing on the PARTS Act, 2 February 2016.

[44]See Autorité de la concurrence 2012.

[45]See section 4.2.1.1.

[46]"Original Equipment Manufacturer" is defined as "one that produces complex equipment (…) from components usually bought from other manufacturers". Aftermarket meanwhile is defined as "the market for parts and accessories used in the repair or enhancement of a product". As Skurka stated in his argument, "(c)argo conversion is defined by Eaton to be the retrofitting of an existing passenger aircraft for a different use by, among other things, adding original equipment doors, assemblies and motors". A "latent ambiguity" existed in the differentiation between OEM and aftermarket. Skurka Aerospace, Inc. v. Eaton Aerospace, L.L.C., Nr. 08 Civ. 1565 (N.D. Ohio, E. Div. Order March 31, 2011), available at: https://www.gpo.gov/fdsys/pkg/USCOURTS-ohnd-1_08-cv-01565/pdf/USCOURTS-ohnd-1_08-cv-01565-3.pdf.

[47]See definition in Article 3 nr. 26 Directive 2007/46/EC of the European Parliament and of the Council of 5 September 2007 establishing a framework for the approval of motor vehicles and their trailers, and of systems, components and separate technical units intended for such vehicles (Framework Directive):

"original parts or equipment' means parts or equipment which are of the same quality as the components used for the assembly of a motor vehicle and which are manufactured according to the

supposed to "match the quality" of the motor vehicle's components.[48] In other words, spare parts are not defined based on the vehicle's origin in terms of the manufacturer's distribution system, but instead in consideration of the quality of the components.[49]

Departure from the concept of origin in defining parts was intended to break down the OEM–non-OEM dichotomy that characterized previous terminology and to open up new marketing opportunities for the independent aftermarket. Nonetheless, many consumers still think in terms of original versus non-original parts, with original parts being generally associated with higher quality and higher prices. On occasion, courts have also adopted the view that "aftermarket parts are not of like kind and quality to OEM parts".[50] Eliminating the familiar OEM–non-OEM dichotomy risks creating the perception that non-OEM parts are the legal equivalent of OEM parts and results in uncertainty with respect to a question that is critical to the consumer.

Both consumers and OEMs would benefit if this gap were filled. Consumers would have access to more information about the origin and quality of spare parts and be enabled to make more informed choices. OEMs would be benefited by the general perception that OEM sourced products are safer. Clear marking laws would facilitate forensic investigation of the origin of parts in capital accidents and long-term would result in fewer substandard parts being installed.

(Footnote 47 continued)

specifications and production standards provided by the vehicle manufacturer for the production of components or spare parts for the motor vehicle in question. This includes spare parts which are manufactured on the same production line as these components. It is presumed, unless the contrary is proven, that parts constitute original spare parts if the part manufacturer certifies that the parts match the quality of the components used for the assembly of the vehicle in question and have been manufactured according to the specifications and production standards of the vehicle manufacturer". See also Article 1(1)(a) of former Commission Regulation (EC) No 1400/2002 of 31 July 2002 on the application of Article 81(3) of the Treaty to categories of vertical agreements and concerted practices in the motor vehicle sector. However, this Regulation was superseded by the Regulation No 461/2010 on the application of Article 101(3) of the Treaty on the Functioning of the European Union to categories of vertical agreements and concerted practices in the motor vehicle sector, which does not contain the definition.

[48]Cf. Article 1(1)(u) of former Commission Regulation (EC) No 1400/2002:

"'spare parts of matching quality' means exclusively spare parts made by any undertaking which can certify at any moment that the parts in question match the quality of the components which are or were used for the assembly of the motor vehicles in question". However, the superseding Regulation (EC) No 461/2010 does not contain the definition.

[49]See International Federation of Automotive Aftermarket Distributors (FIGIEFA), The new Automotive Block Exemption Regulation 1400/2002/EC. Opportunities for Independent Automotive Parts Distributors & Other Independent Aftermarket Operators, available at: http://www.aftermarket.ch/fr/docsDownload/GVO_Brosch_Figiefa_en.pdf.

[50]See Smith v. Am. Family Mut. Ins. Co., 289 S.W.3d 675 (Mo. App. 2009), holding that in using aftermarket parts, the defendant insurance company had breached its agreement with its policy-holder, which required it to return the damaged vehicle to pre-loss condition. Settlement of class action for Farmers Insurance for failure to use certified Certified Automobile Parts Association (CAPA) parts. See Lebrilla v. Farmers, 119 Cal. App. 4th 1070 (June 2004).

6.4.2 Disclosure Models

Disclosure laws have been implemented by national and local governments in many countries in various forms, whether as part of a repair clause or separately, including (1) independent provisions requiring labeling and marking of non-OEM spare parts, (2) conditioning effectiveness of a repair clause upon disclosure, or (3) requirements that repair service providers make available origin and quality information at the time of repair, possibly requiring the consumer's consent to use non-OEM parts. Many other configurations of information can be envisioned.[51]

The idea of disclosure is not foreign to legislatures. The original repair clause submitted by the European Commission in 2004 proposed a very general disclosure requirement to the effect that it "shall ensure that consumers are duly informed about the origin of spare parts so that they can make an informed choice between competing spare parts".[52] A subsequent proposal made the effectiveness of the repair clause contingent on disclosure, however, the scope of the disclosure was watered down in this requirement, shifting it to use of trademarks or trade names.[53]

---

[51]The following provides a comprehensive list of provisions that are relevant under the circumstances and whose disclosure would benefit stakeholders in this debate: 1. identification of any parts not manufactured by (based on authorization from) the manufacturer of the motor vehicle in question; 2. whether the part in question is manufactured by (or based on authorization from) the manufacturer of the motor vehicle in question; manufactured on the same production line as the motor vehicle in question; manufactured according to the specifications and production standards provided by the vehicle manufacturer for the production of components or spare parts for the motor vehicle in question; of the same quality as the components used for the assembly of a motor vehicle; tested to the manufacturer's specifications; guaranteed to meet the manufacturer's specifications; certified by a neutral certification agency; new, reconditioned or rebuilt; 3. whether the part is under warranty by the manufacturer of the motor vehicle in question; if not, disclosure of any applicable warranty information; 4. the part manufacturer's name, logo, identifying numbers and country of origin; 5. the country of manufacture of the spare part; 6. any material differences between the spare part at issue and the manufacturer's specifications that would affect safety, e.g. use of non-tempered glass.

[52]European Commission, Proposal for a Directive amending Directive 98/71/EC 2004: "Article 14 1. Protection as a design shall not exist for a design which constitutes a component part of a complex product used within the meaning of Article 12(1) of this Directive, for the purpose of the repair of that complex product so as to restore its original appearance. 2. Member States shall ensure that consumers are duly informed about the origin of spare parts so that they can make an informed choice between competing spare parts".

[53]Position of the European Parliament, adopted at first reading on 12 December 2007 with a view to the adoption of Directive 2008/…/EC of the European Parliament and of the Council amending Directive 98/71/EC on the legal protection of designs, available at: http://www.europarl.europa.eu/sides/getDoc.do?pubRef=-//EP//TEXT+TA+P6-TA-2007-0609+0+DOC+XML+V0//EN: "Article 14 Designs incorporated in component parts used for repair purposes.
1. Protection as a design shall not exist for a design that is incorporated in or applied to a product which constitutes a component part of a complex product and is used within the meaning of Article 12(1) ‖ for the sole purpose of the repair of that complex product so as to restore its original appearance. This provision shall not apply where the primary purpose of putting the aforesaid component part on the market is other than the repair of the complex product.

However, so far, Article 14 Design Directive has not been changed. The Design Regulation is silent on this matter as well (cf. Article 110(1) CDR).

A good model for disclosure laws is provided by U.S. state laws. Most of them have been enacted as separate pieces of legislation meant to create disclosure obligations for the benefit of insurance companies.[54] They might require disclosures such as "the crash parts are supplied by a source other than the manufacturer of the motor vehicle and are not warranted by the manufacturer of the vehicle".[55] Alternatively, repair shops must disclose whether the parts installed are (1) new, by original equipment manufacturer, (2) new, by non-original equipment manufacturer, (3) used, (4) reconditioned, or (5) rebuilt.[56] In some instances, aftermarket parts are required to carry sufficient permanent identification,[57] to have the manufacturer's logo or name affixed and installed visibly,[58] to state the identity of the manufacturer,[59] or to indicate whether the part is certified.

In crafting an effective disclosure provision, certain practical problems must be addressed. One is placement of the disclosure or marking so as to be visible upon inspection, but also in a way to enable forensic investigation in case of destruction of a vehicle,[60] including specifications of type of damage that the marking must

(Footnote 53 continued)

2. Paragraph 1 shall apply provided that consumers are duly informed about the origin of the product used for the repair by the use of a marking, such as a trade mark or a trade name, or in another appropriate form so that they can make an informed choice between competing products offered for use in effecting the repair.

3. Paragraph 1 shall apply only in respect of visible component parts in the after market once the complex product is marketed in the primary market by the holder of the design right therein or with his consent.

Disclosure laws can operate as complements to the use of trademarks: while trademarks prevent consumers from being subject to misleading statements, disclosure laws affirmatively require that information about a product be provided to the user. Along with trademarks, disclosure laws address OEM concerns such as safety, reputation and liability that might arise out of lower quality products. It merely limits the possibility of misleading statements. A further step, an affirmative duty to inform the public is required.

[54]"At least 35 states have enacted statutes or regulations concerning non-OEM aftermarket or reconditioned parts used in vehicle repairs. Legal requirements governing reconditioned and aftermarket components vary by state, but most are a variation of a National Association of Insurance Commissioners (NAIC) model regulation on the subject, according to the report" (see https://www.cga.ct.gov/2009/rpt/2009-R-0398.htm). This type of law could be easily implemented as a separate flanking measure to a repair clause.

[55]http://takingthehill.com/wp-content/uploads/disclosure.pdf.

[56]Colorado Motor Vehicle Repair Act FN see Colorado Vehicle Repair Act (Motor Vehicle Repair Act, Colo, Rev. Stat s. 42-9-107 (2009)).

[57]Georgia, see http://takingthehill.com/wp-content/uploads/disclosure.pdf.

[58]Iowa, see http://takingthehill.com/wp-content/uploads/disclosure.pdf.

[59]West Virginia, see http://takingthehill.com/wp-content/uploads/disclosure.pdf.

[60]In case of a crash, it is important to be able to identify whether the parts are original or not in order to properly allocate responsibility. See also Widmer and Rajan 2016, p. 6.

survive the crash.[61] A further problem is ensuring that origin information is effectively passed down through successive sales to consumers, including direct purchasers of recycled parts, online sales, etc., from the initial distributor or service provider in possession of that information.

In summary, disclosure provisions either incorporated into a repair clause or adopted as a complementary measure would likely increase acceptance of the repair clause by OEMs, as well as by consumer representatives.

6.5 Parts Certification Requirements

6.5.1 General Considerations

Mandatory certification of aftermarket spare parts could constitute an important measure complementary to a possible repair clause, whether as part of the clause itself or as a separate measure.[62]

Conditioning the exemption upon certification of spare parts might make the clause more acceptable to some of the stakeholders. Certification is important for public safety of vehicles on the road.[63] With increasing technological complexity, use of all kinds of sensors in the automobile's outer shell, for purposes of collision avoidance, lane detection, etc., as well as with rapid advancement towards autonomous driving, concerns regarding faulty aftermarket parts that fail to work with a vehicle's electronic system are increasing.[64] This poses serious safety issues which might be addressed by

[61]For instance, in case of fire, specifying the severity of the fire.

[62]For instance, the recommendation of the Romanian Competition Authority to adopt the repair clause, was flanked by a measure that would empower the national technical body (RAR) to certify vehicles for sale or use on roads and, respectively, refuse it for those which present a safety risk. See Romanian Competition Authority, Recommendation 2012; also Europe Economics 2015, p. 142.

[63]As a general proposition, findings exist to the effect that the increased presence of aftermarket parts in the market does not impact safety. In its Proposal for a Directive amending Directive 98/71/EC (2004), the European Commission concluded that adoption of the repair clause has had no impact on vehicle safety or security. The inquiry prepared in view of the amendment, found that none of the vehicle manufacturers reported a decrease in the safety of persons in countries which had exempted spare parts from protection, such as at the time, Italy and the UK. Of course, with the passage of time, broader liberalization and an increasing number of aftermarket parts presently in the market, these conclusions must be revised.

[64]Certain collision parts can seriously impact safety, if they vary from the original specifications in terms of their corrosion resistance, metallurgical properties, performance in accidents, etc. Mostly these characteristics are not apparent through a purely visual inspection. For instance, a fender is designed to absorb energy in a crash, by way of a "crumple zone" of a specified size, achieved by a certain metallurgical quality of the sheet metal used. Varying the material would change the crumple zone and lead to passenger injury, even though the part's outer appearance would remain the same. See Testimony of Don Risley, ASA, at Congressional hearing on the PARTS Act, February 2, 2016. Furthermore, as the technological complexity of automobiles advances, even a

mandatory certification of non-OEM parts. Certification would further give consumers better means to make a confident and informed choice of aftermarket parts.[65] Finally, mandatory certification would create a uniform objective standard and avoid that stakeholders in the spare parts market attempt to influence the selection of parts on grounds other than safety.[66]

These benefits do not come without drawbacks. Implementation of such measures is associated with costs and administrative burdens and may result in reduced competitiveness and profits of independent spare parts manufacturers. Still, if correctly balanced, a measure of this nature might constitute a helpful negotiation tool towards compromise.

6.5.2 *Current Models of Certification*

Practices regarding approval and/or certification of motor vehicles and spare parts vary considerably from country to country. Many countries have established governmental test agencies or have designated privately-owned ones for testing new models and issuing certificates of conformity with the approved type.[67] However, the system is not uniform and some countries lack certification systems altogether.[68]

(Footnote 64 continued)

car's exterior parts are equipped with sensors of all types, such as collision avoidance sensors, lane detection sensors, etc. A car's collision parts are designed and behave as a system during a crash. With autonomous vehicles, the number and complexity of such sensors will further increase. Even where the manufacturer's specifications are available, the complex electronics, crash detection and prevention capabilities may not be easily replicable by non-authorized parts manufacturers (see Huetter 2016).

[65]Europe Economics 2015, p. 142. This is important in light of imprecise terminology used to describe aftermarket parts, such as "original spare parts" and "spare parts of matching quality" which may designate both OEM parts and parts manufactured by third parties. Particularly in online sales, these designations may prove confusing.

[66]Several lawsuits in the U.S. have alleged that company practices of designating their own captive certification agencies potentially encourage low quality, possibly unsafe, aftermarket parts for cost reasons. Such practices caused at least one court to state that "aftermarket parts are not of like kind and quality to OEM parts", given their engineering, production, and materials. See Smith v. Am. Family Mut. Ins. Co., 289 S.W.3d 675 (Mo. App. 2009), involving a class action against American Family Insurance that alleged breach of contract with the policyholders on the ground that American Family Insurance followed a practice of "specification of 'inferior' non-OEM crash parts for repairs". Similarly, Lebrilla v. Farmers, 119 Cal. App. 4th 1070 (June 2004), involved a class action suit that alleged non-original manufacturer parts did not meet the quality standards set by the company's car insurance policies. See also Bronstein and Griffin 2015. While such statements must be treated as anecdotal, they illustrate possible practices with regard to the use of aftermarket spare parts.

[67]See Directive 2007/46/EC of the European Parliament and of the Council of 5 September 2007.

[68]See Romanian Competition Council Report 2012.

In the U.S. for instance,[69] automotive and parts manufacturers self-certify compliance based on the vehicle standards developed by the National Highway Traffic Safety Administration (NHTSA).[70] This agency has compiled massive data about vehicle crashes, fatalities and injuries, sets all relevant standards and has become an international resource for traffic safety research.[71] It does not directly approve or certify compliance of motor vehicles or parts. While various private certification agencies exist, such as Certified Automobile Parts Association (CAPA) and NSF, they are primarily meant for use by insurance companies.[72] CAPA, for instance, certifies the quality of automotive parts used for collision repairs. It tests independent spare parts to assess their functional equivalence to OEM parts and is geared to reduce the overall cost of automotive repairs.[73] However, because of its ties to insurance companies, the validity of CAPA's certification procedures has been questioned.[74]

The EU system is based on government regulatory approval in advance of manufacturing. Several Directives operate under the Framework Directive (Directive 2007/46/EC) as well as the Regulations developed by the United Nations Economic Commission for Europe (UN-ECE).[75] This system requires vehicles to obtain governmental "type approval" for the whole vehicle before new models can be launched.[76] Type-approval under the EU system is performed by independent agencies, government approved to perform which implement the certifications,

[69]Automotive standards are developed at federal level. However, individual states may have their own provisions regarding automotive safety. For instance, a South Dakota bill, which was opposed by auto repair groups and which was ultimately tabled, would have declared aftermarket parts certified by CAPA to be the legal equivalent of original equipment parts. See http://www.searchautoparts.com/abrn/news/legislative-news-collision/debate-over-aftermarket-crash-parts-continues#sthash.

[70]NHTSA was established in 1970 under the authority granted to the U.S. Secretary of Transportation by the National Traffic and Motor Vehicle Safety Act of 1966, to issue safety standards for all motor vehicles.

[71]NHTSA's work in compiling data about vehicle crashes, fatalities and injuries has become an international resource for traffic safety research. According to NHTSA, no other country has a similar database. See also Canis and Lattazio 2014.

[72]See http://www.repairerdrivennews.com/2016/01/27/qa-as-vehicles-crash-tests-continue-to-develop-nsf-capa-describe-standards/.

[73]See http://www.city-data.com/forum/automotive/2162916-difference-between-capa-non-capa-parts.html#ixzz43ahMI1bh.

[74]Consumer Reports tests of CAPA-certified fenders found that "the CAPA seal of approval is no guarantee of quality comparable" to those made by auto companies (https://www.washingtonpost.com/archive/business/1999/03/18/consumer-backers-ties-to-insurers-questioned/0761294f-bc34-4a47-b463-80f99f88ab7e/).

[75]United Nations Economic Commission for Europe (UNECE), Agreement concerning the Adoption of Uniform Technical Prescriptions (Rev. 2), 5 October 1995.

[76]European Community Whole Vehicle Type Approval (ECWVTA) is used to ensure that motor vehicles meet relevant environmental, safety and security standards.

including THATCHAM (UK),[77] TÜV,[78] DEKRA[79] (both Germany) and CENTRO ZARAGOZA (Spain),[80] all of which are gradually expanding their services Europe-wide.

Regulation (EC) No 78/2009 on the type-approval of motor vehicles with regard to the protection of pedestrians and other vulnerable road users[81] specifically covers frontal surfaces, likely including windscreens, lights, bonnets and bumpers. The Regulation addresses the entire structure of a vehicle's frontal protection system, which would impact the safety of pedestrians hit by a vehicle's frontal surfaces.[82] It however does not apply to parts other than those pertaining to the frontal structure, which may leave some room for legislative action in this field.[83]

In addition to tightening the EU requirements discussed above, certification would be effective primarily in countries where no spare part certification is required, and would therefore work best on a national basis. However, the complexity of such means must not be underestimated. It would entail empowering a body to approve spare parts, consistent with pre-set technical standards that would ensure compliance. This authority would be entitled to refuse certification, including sale and use on public roads of parts which present a danger. Whether from a public policy standpoint this expenditure is warranted, would have to be decided by each national legislator.

Still, adoption of such certification requirement might contribute to acceptance of the clause by OEMs. For independent spare parts manufacturers, on the other hand, support of this measure would be a difficult choice to make, as profits resulting from market liberalization might be considerably undermined by costs associated with a certification requirement.

6.6 Scope of the Repair Clause

As set forth in Article 110(1) CDR, the repair clause is comprehensively drawn: it covers spare parts in all industries and refers to any part that is used "for the purpose of the repair of that complex product so as to restore its original appearance". A narrower scope might prove more acceptable to the stakeholders. Some of the

---

[77] http://www.thatcham.org.

[78] www.tuv.com.

[79] www.dekra.de.

[80] http://www.centro-zaragoza.com.

[81] Regulation (EC) No 78/2009 of the European Parliament and of the Council of 14 January 2009 on the type-approval of motor vehicles with regard to the protection of pedestrians and other vulnerable road users, amending Directive 2007/46/EC and repealing Directives 2003/102/EC and 2005/66/EC2003, L 35/1.

[82] Regulation (EC) No 78/2009.

[83] For instance, measures are under way to enhance type approval for trucks and motorcycles, but they are being opposed by FIGIEFA.

obvious variables relate to the question whether the clause should be expressly limited to the automotive industry, and if so, whether the parts covered by it should be specifically identified, or left to be construed by courts.

6.6.1 Limiting the Clause by Industry

Different countries' approaches to the scope of repair clauses have varied. The EU and Australia, for instance, have opted for a liberalization across the board, whereas the U.S. legislative repair clause proposal, while initially applicable to all industries, has gradually been narrowed down and is currently limited to "motor vehicles".

In terms of the impact, the automotive industry would clearly be more impacted than others, given its well-developed and economically attractive aftermarket for collision parts.[84] As noted above, this is likely attributable to factors such as the high volume of spare part use—both as a result of normal wear and collisions, as well as the high purchase price of vehicles which mandates repair as opposed to replacement.[85]

In theory, aftermarkets for spare parts should display similar dynamics in other industries, in which interoperability considerations and high switching costs create a lock-in effect that results in users becoming captives to the supplier of the original product for the product's lifecycle.[86] When the only design option is foreclosed by design protection, a competitive market may have difficulties to emerge.

However, evidence of a possible aftermarket in other industries has been sporadic and inconsistent. Occasional lawsuits have alleged anticompetitive behavior in such markets.[87] As mentioned earlier, the EPEC Report refers to a proceeding filed in front of the EU competition authorities alleging competition violations by the watch industry[88] but no statements could be obtained from the watch industry. The fact that so far in none of the other industries the spare part problem has displayed the same persistence as in the auto industry, may be due to an overall lack of attractiveness of a secondary market in the respective sector, including for reasons of low demand for spare parts, limited installed bases of complex products, breakage due to wear alone and to the lesser incidence of no-probability events

[84]See Berns 2013, pp. 1 et seq.

[85]Compared to other lower priced products which tend to be readily replaceable. Separately, but maybe most importantly, collision damage is typically covered by insurance. This fact turns the insurance industry into a stakeholder in the debate, and involves its not inconsiderable lobbying strength.

[86]See van den Bergh and Camasca 2001, pp. 113 et seq.

[87]See Suprema Corte di Cassazione (Italy), 13 May 2014, Case Nr. 37451/2014 – Vorwerk Folletto; Dyson Ltd v. Qualtex (UK) Ltd [2006] EWCA Civ 166 (both cases related to vacuum cleaners); to these, see Hartwig 2016, p. 108.

[88]Europe Economics 2015.

(accidents), the high cost of gearing up for the production of replicated parts, particularly if technologically complex.

Even if evidence of aftermarkets outside the automotive industry is thin to nonexistent, their emergence in the future cannot be excluded. From a policy perspective, a narrow repair clause might foreclose development of this sector unwittingly; whereas a broad exemption might stimulate competition in new sectors.

From a practical perspective, the decision whether to include other industries may well be neutral to stakeholders, as no lobby is to be expected from these industries. The fact that the automotive industry seems to favor inclusion of other industries would indicate that their inclusion might ease negotiation, even if only to a small degree.

6.6.2 Limiting the Clause by Product Type

A further question relating to the scope of the repair clause is whether the parts exempted should be specifically listed in the provision. This latter option was considered in the Explanatory Memorandum of the European Commission's 2004 proposal for amendment of the Design Directive, which limited the clause to body parts, lighting and glass (collectively "crash parts").[89] Similarly, the proposed U.S. bill, lists, by way of example, specific parts that are covered by the clause.[90]

On the one hand, the more specific the list, the more clearly the legislative intent can be captured. On the other, from a practical perspective, an itemized list might be virtually impossible to implement if the clause is not limited by industry.[91] Even if the clause were limited to "motor vehicles", as in the U.S. bill, such itemization would yield countless visible parts that pertain to automobiles, motorcycles, trucks, etc. Apart from the practical difficulties of listing a large number of parts, it may not be realistic to hope for a consensus among stakeholders on specific parts to include.

By default, the question remains to be resolved by the courts. As noted above, the courts will approach the issue by conducting in each individual case an examination on the merits. Its aim is dual: to determine whether the parts are capable of giving rise to monopolization of the spare parts market[92] and to verify whether the application of the clause in the specific instance would yield

[89]See Deutsche Vereinigung für gewerblichen Rechtsschutz und Urheberrecht (GRUR) 2015, citing European Commission, Proposal for a Directive amending Directive 98/71/EC 2004, p. 6.

[90]The U.S. legislative proposal, U.S. PARTS Act, HR 2035, Section 2(A) provides "the term 'component part' — (i) means a component part of the exterior of a motor vehicle only, such as a hood, fender, tail light, side mirror, or quarter panel".

[91]And even if for instance the limitation is to motor vehicles, then trucks, motorcycles, etc. would still be included.

[92]LG Düsseldorf, 28 November 2013, 14 c O 304/12 U, GRUR-RS 2014, 00345.

appreciable economic benefit.[93] So far this has resulted in a narrowing of the scope of the clause by excluding wheel caps and rims and by creating the conceptual framework for a further narrowing. The basis for further narrowing the notion that visible spare parts are "divisible" into parts that are objectively necessary to restore the original appearance of a complex product, and parts which are not and that parts that constitute discretionary upgrades do not meet the requirement of Article 110(1) CDR.[94] In a market of increasing customization of vehicles, this interpretation of the scope of Article 110(1) CDR may end up diminishing the market impacted by the effect of Article 110(1) CDR. This trend in the interpretation of the repair clause may also alter the stakeholders' position vis-à-vis compromise.

6.7 Facilitation of Criminal Enforcement

A further measure that might serve as a bargaining chip in the adoption of a repair clause is facilitation of criminal enforcement for spare part counterfeits.

6.7.1 Criminal Enforcement Methods of IP Rights

Enactment of a repair clause would deprive automotive manufacturers of enforcement means against counterfeit parts. This means that OEMs would have to rely on other IP laws instead, primarily on trademark laws, including on the criminal enforcement provisions of such laws.[95] Strengthening the ability to rely on criminal enforcement would provide OEM's with an alternative avenue for fighting counterfeits. The following will examine the availability of such measures, as well as some of the practical issues that arise in their application.

6.7.2 Foundation in International Law

While not accepted by all jurisdictions, criminal enforcement is increasingly becoming popular among rightholders. Since the adoption of TRIPS, availability of criminal procedures has become mandatory "at least in cases of willful trademark

---

[93]LG Düsseldorf, 28 November 2013, 14 c O 304/12 U, GRUR-RS 2014, 00345.

[94]BMW AG v. Round and Metal Ltd and Philip David Gross [2012] EWHC 2099 (pat): OLG Stuttgart, 11 September 2014, 2 U 46/14, GRUR 2015, 380; for a more detailed discussion on the cases, see supra section 5.2.

[95]E.g. in Germany §§ 143 et seq. MarkenG; for details see Kaiser, in: Ambs and Häberle 2015, MarkenG, §§ 143 et seq.

counterfeiting or copyright piracy on a commercial scale".[96] Remedies include imprisonment, monetary fines sufficient to constitute a deterrent, as well as in appropriate cases, the seizure, forfeiture and destruction of the infringing goods.[97] Optionally, criminal procedures are available in connection with other intellectual property rights as well, for violations that are willful and on a commercial scale. Accordingly, all TRIPS member states are required to provide criminal measures for trademark infringement. The availability of such remedies for design protection will vary among jurisdictions.

6.7.3 IP-Specific Criminal Enforcement at European Level

In Europe, the harmonized enactment of criminal enforcement of IP laws is limited. The currently effective Directive 2004/48/EC on the enforcement of intellectual property rights (Enforcement Directive)[98] merely acknowledges the existence of national criminal IP law enforcement laws (see e.g. Recital 28 and Article 2(3)(c)) but contains no specific substantive provisions.

To fill this gap and to bring about harmonization,[99] in 2006 the European Commission had proposed a Directive on criminal measures aimed at the enforcement of intellectual property right.[100] However, the proposal was withdrawn in 2010 and therefore the Directive was never enacted.[101] This outcome was not perceived as a loss, as the proposal was problematic in several regards, in particular as it related to design protection of spare parts. First, the proposed Directive would have conflicted with the contemplated liberalization of the spare parts market through relaxation of design protection, a goal which the EU legislator had been pursuing separately.[102] Second, Article 3 of the proposed Directive would have raised "intentional infringement(s)" of IP rights to the level of criminal offenses. This broad formulation of the Commission's 2006 proposal, intended to address counterfeits, would also have captured certain protected spare parts and their

---

[96] Article 61s. 1 TRIPS.

[97] Article 61s. 2, 3 TRIPS.

[98] Directive 2004/48/EC of the European Parliament and of the Council of 29 April 2004 on the enforcement of intellectual property rights, L 195/16.

[99] Drexl 2012, pp. 370 et seq., also for details on the legislative power of the EU in this field.

[100] European Commission, Amended proposal for a Directive of the European Parliament and of the Council on criminal measures aimed at ensuring the enforcement of intellectual property rights of 26 April 2006, COM(2006) 0168 final. The comments in this section refer to the version including the amendments proposed by the European Parliament (see Resolution of the European Parliament of 25 April 2007, Enforcement of intellectual property rights (criminal measures), P6_TA(2007)0145)). See also Drexl 2012, pp. 370 et seq.; Hilty et al. 2006, pp. 970 et seq.

[101] European Commission, Withdrawal of obsolete Commission Proposals, C 252/7, 18 September 2010, p. 3.

[102] See Article 14 Design Directive; for this, see also supra section 4.2.1.1.

importation from member states which had adopted the repair clause, into member states which had not.[103] A 2007 amendment to the proposal mitigated this effect somewhat by excluding parallel imports.[104] Thus, at least imports of goods placed in circulation in member states which had adopted the clause, into member states which had not, would be exempt from criminal liability.[105] Still the formulation remained broad. Finally, the prohibition of infringement on a "commercial scale" (Article 3) risks being equated with "any infringement of an intellectual property right committed to obtain a commercial advantage" (Article 2(b)). This would have subjected any repair shop's commercially motivated spare part installation to criminal liability, unless authorized by the rightholder.[106]

In short, for practical purposes, the EU lacks a harmonized criminal IP infringement enforcement mechanism and enforcement must therefore be premised on national laws.

6.7.4 IP-Specific Criminal Enforcement at National Level

National criminal law may vary from one jurisdiction to another. However, as a practical matter, even though not harmonized, criminal enforcement in the IP field is likely to follow the general tenor of Article 61 TRIPS, at least as far as substantive norms are concerned. Most countries provide criminal IP enforcement under national laws, such as in Germany §§ 143 et seq., MarkenG; § 51 DesignG; in France, Article L615-12 Code de la propriété intellectuelle; in Italy, Article 127 CPI; in Poland, Articles 303 et seq. The penalties imposed vary from mere misdemeanors to felonies. There is some pressure to increase penalties to combat counterfeits.[107] The actual application of these laws however, is likely to vary: in some EU member states, including France and Poland, the penal aspect of IP right infringement actions tends to be examined with great thoroughness, whereas, in Germany, prosecutors and courts tend to be both reluctant and slow when it comes to criminal prosecution of IP infringements.

---

[103]Cf. Drexl 2012, p. 382.

[104]Drexl 2012, p. 383.

[105]See also Drexl 2012, p. 383.

[106]Drexl 2012, p. 383.

[107]For instance, the Romanian Competition Council proposed that the sale of counterfeit automotive products utilized be raised to the level of a felony (Romanian Competition Council, Report, 2012).

6.7.5 Practical Issues Relating to Criminal IP Enforcement

One of the reasons for which in countries such as Germany criminal enforcement proceedings are used relatively infrequently, may be the prosecutors' and judges' unfamiliarity with these laws and consequent lack of interest in handling such cases. Indeed, the handling prosecutor or judge who is assigned a registered trademark or design, must first evaluate the validity of the respective type of IP, under the respective requirements of substantive law.[108] IP law is generally not a mandatory subject in law studies,[109] and thus professionals who are not specialized in the field often have difficulties in making decisions. Both determination of validity and comparison of the similarity of the protected and the infringing mark or design, are extremely complex and fraught with uncertainties. Not surprisingly, prosecutors are not eager to take on these matters.

In short, improvement of enforcement measures in other areas of IP law, in particular in trademark law, would remove pressure from design law. Steps towards ensuring more effective criminal enforcement could act as flanking measures to a compromise solution on the repair clause. For instance, continuing education for prosecutors and judges in enforcement of IP laws might be mandated in EU member states. Selected IP courses would be more effective than say introducing IP as a mandatory subject matter, as they would be closer to practice both in terms of time and of content. Alternatively, selected prosecutors and judges might receive additional training in these areas of law, so that they could serve as resources for their colleagues. Thus, imposing certain concrete organizational measures in other areas of IP law, to accompany a compromise repair clause, could constitute a means of better enforcing existing law and compensating for the effect of the repair clause.

6.8 Remunerated License

6.8.1 Introduction

A repair clause would meet fewer objections, of both doctrinal and economic nature, if a remuneration provision were included. While spare parts would remain under design protection, their use would be subject to royalty payments to the original manufacturer.

A well-working remuneration provision has the advantage of leveling the playing field: it would provide (some) compensation to the manufacturer/designer,

---

[108]As required for instance, by § 51 DesignG. For a detailed treatment see Eichmann, in: Eichmann et al. 2015, § 51.

[109]As an example, see Hamburger Verordnung über Prüfungsgegenstände der staatlichen Pflichtfachprüfung im Rahmen der ersten Prüfung (Prüfungsgegenständeverordnung) of 23 December 2003, available at: http://justiz.hamburg.de/1-examen/.

allowing spare part manufacturers lawful access to protected parts, and reducing consumer prices. This solution would be viewed favorably by courts, who have based their narrow interpretation of the repair clause in part, on its uncompensated nature.[110]

6.8.2 *History*

The idea of a remuneration option is not new. The European Commission's Amended Proposal for a Design Directive of 1996 contained this type of provision. It was never acted upon, in large part because no agreement could be reached on the basis for calculating remuneration.[111] Furthermore, the provision was considered administratively too burdensome and was abandoned.[112] However, the issue remains open.

Greece was the only country which adopted the European Commission's early proposal in 1997 essentially verbatim. The Greek repair clause provided for a five-year royalty-free protection period for spare parts, followed by payment of "fair and reasonable remuneration to the owner as agreed by the parties or in case of dispute, as decided by the competent Court".[113] Apparently the remuneration system has never been implemented, due to an inability to agree on the amount of royalties.[114]

6.8.3 *Legal Basis for a Remunerated License Solution*

Legal justifications for a remunerated license can be easily provided under both IP law and a competition law approaches.

From an IP perspective, a remunerated license would likely eliminate the objection under Article 26(2) TRIPS.[115] An uncompensated measure would be perceived as exceeding the "limited exceptions" allowable by TRIPS and the loss of income as causing unreasonable prejudice to the rightholder's legitimate interests.[116] In contrast, compensation would likely limit the loss to a level, which,

[110]See LG Hamburg, 18 September 2015, 308 O 143/14, GRUR-RS 2015, 16872; LG Düsseldorf, 28 November 2013, 14 c O 304/12 U, GRUR-RS 2014, 00345.

[111]Ohlgart, in: Franzosi 1996, p. 149.

[112]See Europe Economics 2015; Autorité de la concurrence 2012.

[113]For text, see supra section 6.2.2.

[114]Europe Economics 2015, p. 142.

[115]To this, see supra section 4.1.2.1.

[116]"(P)rejudice to the legitimate interests reaches an unreasonable level if it causes an unreasonable loss of income" (WTO 1999, para 6.229); see also Straus 2005, pp. 965 et seq.

depending on the specific circumstances, might not be deemed unreasonable.[117] Scholars tend to agree that a repair clause combined with a compensation provision is less likely to violate a design owner's property rights, as the loss of exclusivity is, to some extent, compensated.[118] The same reasoning may well remove constitutionally based legal objections.[119]

Alternatively, justification for a remunerated license could be founded in competition law. As part of the general market regulation framework, EU laws provide means for correcting possible unjustified restriction of competition in the secondary market as a result of design patent protection.[120] Precedential rulings on point are not available, as competition authorities have declined to take action in connection with the secondary market for spare parts. But other areas of IP might serve as guidance. For instance, the area of standard essential patents displays certain commonalities with the spare parts aftermarket, in that no realistic practical alternative to the protected rights exists.[121] The holder of a patent, which is essential to the standard, is required to grant a license on terms that are fair, reasonable and non-discriminating.[122] Therefore, at least in theory, doctrinal support for a compensated license can probably be found in EU competition law.[123] However, as discussed above, applicable CJEU jurisprudence found no abuse of a dominant position in the automotive spare parts market and therefore does not mandate

---

[117]WTO 1999.

[118]Kur 1996, p. 886; Straus 2005, n. 62.

[119]See supra section 4.1.2.

[120]Ullrich (2015), pp. 377 et seq. These cases dealing with license denials may serve as some guidance on possible justifications for a remunerated license, are however not necessarily applicable to the present issue. While refusal to license per se is not a violation of competition law, it may become so in the event of the following: (1) the refusal to license prevents the emergence of a new product, for which consumer demand exists; (2) no objective consideration can justify the refusal; and (3) the refusal provides the intellectual property owner with monopoly in the secondary markets of its intellectual property. See CJEU, 29 April 2004, Case C-418/01 – IMS Health GmbH & Co. OHG v. NDC Health GmbH & Co. KG [2004] I-5039; CJEU, 6 April 1995, Joined cases C-241/91 P and C-242/91 P – Radio Telefis Eireann (RTE) and Independent Television Publications Ltd (ITP) v. Commission of the European Communities (Magill) [1995] I-743; European Commission, 24 March 2004, Case COMP/C-3/37.792 – Microsoft, relating to a proceeding under Article 82 of the EC Treaty; but see Suthersanen 2000, para 10-006: abuse requires an act beyond the normal exercise of the right "only conduct which exceeds the subject matter will be condemned".

[121]Lianos and Dreyfuss 2013.

[122]Maume (2015), pp. 95 et seq.; see also CJEU, 16 July 2015, Case C-170/13 – Huawei Technologies Co. Ltd v. ZTE Corp. and ZTE Deutschland GmbH, GRUR 2015, 764; BGH, 6 May 2009, KZR 39/06, BGHZ 180, 312 – Orange-Book-Standard (for Germany).

[123]E.g. CJEU, 29 April 2004, Case C-418/01 – IMS Health GmbH & Co. OHG v. NDC Health GmbH & Co. KG [2004] I-5039; CJEU, 6 April 1995, Joined cases C-241/91 P and C-242/91 P – Radio Telefis Eireann (RTE) and Independent Television Publications Ltd (ITP) v. Commission of the European Communities (Magill) [1995] I-743.

non-voluntary license.[124] A compensated license solution, whether legislated or not, would therefore not be predicated on the mandates of EU competition law, but rather be based on a voluntary agreement among stakeholders as part of a compromise solution.

6.8.4 Implementation of a Remunerated License

Separate from the legal aspect, the practical hurdles in structuring a negotiated agreed-upon remunerated license are immediately apparent. These relate primarily to determining the correct level of remuneration and finding the administrative means of implementing the measure.[125] The measure has already once failed due to such obstacles. However, it must be kept in mind that developmental leaps have taken place in both law and technology since the time this issue has last been examined in depth in the 1990s at the time the Design Directive was debated, and that new means may be available.

In terms of setting royalties, some basic guiding principles in determining methods of remuneration have been firmed up. Thus in the case of spare parts, it is clear that the OEM's remuneration the purpose of a license is purely compensatory in nature, meant to correct a market irregularity. Further, remuneration should be set based on arm's length dealings in a competitive market. No social goals are pursued by such license; they simply serve market regulation.[126] Finally, the stakeholders' interests must be balanced so as to create appropriate conditions for spare parts manufacturers to compete in the market, without eviscerating the OEM's IP rights.

The law on calculation of royalties has advanced and provides numerous options.[127] In the context of spare parts, the most realistic basis seems to relate to the costs incurred by the original producer in developing the design.[128] A detailed

---

[124]Cf. e.g. CJEU, 6 April 1995, Joined cases C-241/91 P and C-242/91 P – Radio Telefis Eireann (RTE) and Independent Television Publications Ltd (ITP) v. Commission of the European Communities (Magill) [1995] I-743; CJEU, 5 October 1988, Case C-238/87 – AB Volvo v. Erik Veng (UK) Ltd. [1988] 6211; for a more detailed discussion, see supra section 4.1.2.2.

[125]European Commission, Proposal for a Directive amending Directive 98/71/EC 2004.

[126]See Lin (2015), pp. 165 et seq., also setting forth various methods of determining reasonable royalties or adequate compensation, primarily in pharma.

[127]E.g. the hypothetical negotiation method, assuming a willing licensor and license, based on 15 factors. See Georgia-Pacific Corp. v. United States Plywood Corp., 318 F. Supp. 1116 (S.D.N.Y. 1970), mod. and aff'd, 446 F.2d 295 (2d Cir. 1971), cert. denied, 404 U.S. 870 (1971); percentage of cost savings achieved by the patented method Hanson v. Alpine Valley Ski Area, Inc., 718 F. 2d. 1075 (Fed. Cir. 1983); calculation based on the projected profit from infringing items TWM Manufacturing Co. v. Dura Corp., 789 F. 2d 895 (Fed. Cir. 1986), cert. denied, 479 U.S. 852 (1986).

[128]Ohlgart, in: Franzosi 1996, p. 149.

formulation of the appropriate basis for remuneration would be required, presumably on the premise that the OEMs' investment could be used as a starting point.[129]

The Commission's originally articulated standards for calculating a remuneration that is "fair and reasonable" may still apply, possibly with the addition of "non-discriminatory". Guidance on the application of these standards can be derived from the wealth of cases primarily in the patent field and from scholarship relating to determination of royalties in patent litigation.[130] New uses have emerged in the context of standard setting and patent pools which heavily turn on determination of reasonable royalties. A wealth of empiric data on royalty models and calculations has also become available.[131]

When it comes to setting royalty rates in a market of mass produced high volume products, transaction costs can easily outweigh the benefit to the stakeholders. To minimize transaction costs, royalty rates must not be set by courts, as in FRAND cases, but by the legislature. Furthermore, mechanisms must be in place to minimize court intervention.[132] In the analogous setting of standard essential patents and the FRAND commitments governing them, Lemley and Shapiro envision this scenario as follows.[133] The authors propose individually negotiated licenses based on an *ex ante* FRAND commitment, agreement could be based on a monetary offer to take a license made by the putative licensee to the rightholder, whereby the rightholder is required to reciprocate. If no agreement can be reached, the question is submitted to simplified binding arbitration, which precludes court actions. This solution may appear cumbersome in the context of spare parts if handled by traditional means. However, if this idea is implemented electronically, the logistical hurdles might be overcome. An online platform could be structured so as to support

---

[129]Ohlgart, in: Franzosi 1996, p. 149.

[130]Some of the main royalty computation methods proposed are (1) the hypothetical negotiation method (while Georgia-Pacific in principle remain the main governing authority, more recent cases have provided adjustments; e.g. Uniloc USA, Inc. v. Microsoft Corp., 632 F.3d 1292 (Fed. Cir. 2011), Ericsson, Inc., et al. v. D-Link Systems, Inc., et al. (Fed. Cir. 2014) (suggesting that courts do not need to rely on every one of the 15 Georgia-Pacific factors and instead focus only on the ones relevant in a FRAND context)); (2) computation of royalties based on a set of identified comparable licenses (3) the incremental value approach, which sets royalties based on the incremental value of the patented technology over the next-best alternative; (4) the *ex ante* approach which considers that cost of the function performed by the patent, prior to the patent becoming a SEP; (5) the "top-down" approach calculates per unit profits and computes the royalty based on the total number of SEPs in the standard, In re Innovatio IP Ventures, LLC Patent Litigation, No. 11, C-9308, 2013 WL 5593609 (ND Ill., 3 October 2013). For the EU, see CJEU, 16 July 2015, Case C-170/13 – Huawei Technologies Co. Ltd v. ZTE Corp. and ZTE Deutschland GmbH, GRUR 2015, 764.

[131]E.g. PricewaterhouseCoopers (PwC), Patent Litigation Study. As case volume leaps, damages continue general decline, 2014, available at: http://www.pwc.com/us/en/forensic-services/publications/2014-patent-litigation-study.html.

[132]Ullrich, 2015, pp. 377 et seq.: if no pre-established royalty can be agreed upon, "legislator may leave fixing of the remuneration to the parties. Just to be reviewed by a court in accordance with standards set by the country, usually vague, under Article 31 TRIPS".

[133]Lemley and Shapiro 2013, pp. 1135 et seq.

a negotiation or bidding process within a pre-established range. The successful bid would then be tracked, electronic payment effectuated and submitted to some form of simplified dispute resolution only in case of non-compliance. A platform of this nature would also address concerns, such as informing the rightholder about the intended use and providing information relating to the volume of use, which the Commission perceived to constitute a stumbling block in its 1996 proposal for a remunerated repair clause.[134] While this type of solution would primarily address the transaction side, it could also be coupled with the enforcement side, by for instance, legislatively limiting sales by independent distributors or service providers to parts that have been validated by this system.

Technologies required to implement such a solution exist and have been applied in many lucrative areas such as online sales, online auctions and music royalty collection. The notion of a remunerated license has thus become less of a legal or policy problem than a technical one. In the digital age, a technical problem of this nature should be solvable if the will exists. A relatively gentle push from the European Commission or other executive authority might be all that is required to implement this type of solution.

Finally, with the advancement of 3D printing technology, it is anticipated that licensing relations between OEMs and independents will undergo a fundamental transformation. This aspect will be discussed in detail in the next chapter.

This concludes the section discussing possibly compromise-furthering considerations. The aspects addressed are intended to be merely illustrative. Given the complexity of the spare parts business, there are many more aspects that tend to favor one or another of the stakeholders, a circumstance that would lend itself to a trading of benefits.[135] Creativity in devising solutions, political will and an

---

[134]Such administrative conditions are necessary to conduct a successful licensing program, including informing the rightholder about the intended use and offering reasonable remuneration, and providing regular and reliable information relating to the volume of use. See European Commission, 31 January 1994, COM(93) 344 final, ABl. EG C 29/32; also Berns 2013, note 1969.

[135]Other legislative options could serve to shift the balance between stakeholders and facilitate a compromise. One area of possible leverage relates, for instance, to the insurance industries' increasing power in the aftermarket, which correlates with consumer- and OEM-unfriendly behavior. Automotive insurers tend to partner with selected repair services, in exchange for promises of reduced prices on non-original spare parts. It has become a common practice for insurers to "steer" consumers towards their partner repair services. See Doggett 2016, pp. 2 et seq. This practice can be addressed by prohibiting insurers from "steering" their customers to repair services that use non-original spare parts. Such anti-steering laws are well-known to U.S. state law (cf. Doggett 2016, pp. 3 et seq.) Such laws would favor the automotive industry. It is likely that a segment of consumers, given the choice, would be inclined to opt for repair services that use original spare parts. Somewhat comparable results can be obtained by requiring independent repair services to use original spare parts when exchanging parts for repair or other purposes, at least within the warranty period. Another flanking measure specifically addressing the safety concerns relating to the use of non-original spare parts would be to require independent repair services with genuine access to OEM technical information, allowing them to create parts of the same quality and safety standards (Autorité de la concurrence 2012, para 242 et seq.). However, this would weaken the OEM's position and it is therefore unlikely to be accepted by them in a compromise.

authority with the necessary power to elicit concessions at a multinational level from the players would be required. That may be too tall an order to fill.

References

Ambs F, Häberle M (2015) Erbs/Kohlhaas—Strafrechtliche Nebengesetze mit Straf- und Bußgeldvorschriften des Wirtschafts- und Verwaltungsrechts, 204th edn. München, C.H, Beck

Autorité de la concurrence, Avis n° 12-A-21 du 8 octobre 2012 relatif au fonctionnement concurrentiel des secteurs de la réparation et de l'entretien de véhicules et de la fabrication et de la distribution de pièces de rechange (The French Competition Authority's report of 2012). http://www.autoritedelaconcurrence.fr/pdf/avis/12a21.pdf

Berns E (2013) Marktmissbrauch auf Ersatzteilmärkten im deutschen, europäischen und US-amerikanischen Markt [Diss.]. EUL Verlag, Lohmar

Bronstein S, Griffin D (2015) Auto insurers accused of pushing cheap and sometimes dangerous repairs. http://www.cnn.com/2015/02/11/us/auto-repair

Canis B, Lattazio RK (2014) U.S. and EU motor vehicle standards: issues for transatlantic trade negotiations. https://www.google.de/webhp?sourceid=chrome-instant&ion=1&espv=2&ie=UTF-8#

Deutsche Vereinigung für gewerblichen Rechtsschutz und Urheberrecht (GRUR) (2015) Stellungnahme des GRUR Fachausschusses für Designrecht unter Mitwirkung von Mitgliedern des Fachausschusses für Wettbewerbs- und Markenrecht zum Vorabentscheidungsersuchen des Tribunale Ordinario di Torino—Ford Motor Company v. Wheeltrims SRL (Case C-500/14), 13 February 2015. http://www.grur.org/uploads/tx_gstatement/2015-02-13-GRUR-Stn-Vorabentscheidungsersuchen-Tribunale-Torino.pdf

Doggett RK (2016) The automotive-repair industry and the practice of insurance steering: is there a fulfilling escape route available for auto-repair facilities? Southern University L R 1-34

Drexl J (2012) Counterfeiting and the spare parts issue. In: Geiger Christophe (ed) Criminal enforcement and intellectual property: a handbook of contemporary research. Edward Elgar Publishing, Cheltenham (UK), Northampton, MA (USA), pp 369–385

Drexl J, Hilty R, Kur, A (2005a) Design protection for spare parts and the Commission's proposal for a repairs clause, IIC 36:448–457

Drexl J, Hilty R, Kur, A (2005b) Designschutz für Ersatzteile—Der Kommissionsvorschlag zur Einführung einer Reparaturklausel. GRUR Int 449–457

Eichmann H, Vogel von Falckenstein R, Kühne M (eds) (2015) Geschmacksmustergesetz. Kommentar, 5th edn. C.H. Beck, München

Europe Economics (2015) The economic review of industrial rights in Europe—final report, MARKT/2013/064//D2ST/OP, January 2015, (commissioned by the European Commission). http://ec.europa.eu/growth/tools-databases/newsroom/cf/itemdetail.cfm?item_id=8844

European Commission (2004) Proposal for a directive of the European parliament and of the council amending directive 98/71/EC on the legal protection of designs (COM(2004)) 582 final—2004/0203 (COD). http://www.eumonitor.eu/9353000/1/j9vvik7m1c3gyxp/vi7jgsymsdz3

European Communities Trademark Association (ECTA), Design Committee (2016) Position paper designs and spare parts. http://www.ecta.org/IMG/pdf/ECTA_-_Position_Paper_-_Spare_Parts_(003).pdf

Franzosi M (ed) (1996) European design protection. Commentary to directive and regulation proposals. Kluwer Law International, The Hague, London, Boston

Hartwig H (2016) Spare parts under European design and trade mark law. GRUR Int 102–108

Hilty RM, Kur A, Peukert A (2006) Statement of the Max Planck institute for intellectual property, competition and tax law on the proposal for a directive of the European parliament and of the

council on criminal measures aimed at ensuring the enforcement of intellectual property rights. IIC 37:970–977
Huetter J (2016) As vehicles, crash tests continue to develop, NSF, CAPA describe standards. In: Repairer driven news. http://www.repairerdrivennews.com/2016/01/27/qa-as-vehicles-crash-tests-continue-to-develop-nsf-capa-describe-standards/
Lemley MA, Shapiro C (2013) A simple approach to setting reasonable royalties for standard essential patents. Berkeley Technol L J 28:1135–1166
Lianos I, Dreyfuss R (2013) New challenges in the intersection of intellectual property rights with competition law. Centre for Law, Economics and Society (CLES) Working Paper Series 4. https://www.ucl.ac.uk/cles/research-paper-series/research-papers/cles-4-2013
Lin X (2015) Prior negotiation and remuneration for patent compulsory licensing: practice, problem, and proposal. Practical experiences and ways forward. In: Hilty RM, Liu KC (eds) Compulsory Licensing. Springer, Berlin, pp 165–190 In: Germany
Maume P (2015) Compulsory licensing in Germany. In: Hilty RM, Liu KC (eds) Compulsory licensing: practical experiences and ways forward. Springer, Berlin, Heidelberg, pp 95–120
Straus J (2005) Ende des Geschmacksmusterschutzes für Ersatzteile in Europa?. Das Mandat der Kommission und seine zweifelhafte Ausführung. GRUR Int, Vorgeschlagene Änderungen der EU-Richtlinie, pp 965–979
Ullrich H (2015) Mandatory licensing under patent law and competition law: different concerns, complementary roles. In: Hilty RM, Liu KC (eds) Compulsory licensing: practical experiences and ways forward. Springer, Berlin, Heidelberg, pp 377–396
Van den Bergh R, Camasasca P (2001) European Competition Law and Economics. A Comparative Perspective, Intersentia, Antwerpen (Belgium)
Widmer M, Rajan V (2016) 3D opportunity for intellectual property risk. Additive manufacturing stakes its claim. https://dupress.deloitte.com/content/dam/dup-us-en/articles/3d-printing-intellectual-property-risks/ER_2981-3D-opportunity-intellectual-property_MASTER.pdf
World Trade Organization (WTO) (1999) Dispute panel report on section 110(5) of the U.S. Copyright Act, WT/DS160/5, April 16, 1999. https://www.wto.org/english/tratop_e/dispu_e/cases_e/ds160_e.htm
Zwanzger S (2010) Das Gemeinschaftsgeschmacksmuster zwischen Gemeinschaftsrecht und nationalem Recht. Zugleich ein Beitrag zur europäischen Zivilrechtsdogmatik, Mohr Siebeck, Tübingen

Chapter 7
Future Outlook: The Spare Parts Debate in the Era of 3D Printing

7.1 Introduction

In the preceding chapters we examined the law and policy surrounding the issue of design protection for spare parts. We found it to be deeply divided, a reflection of the stakeholders' respective interests in the market. The situation has been stagnant but increasingly tense for the past two decades, and while legislative action remains possible, the will for it seems to be lacking. Solutions are being sought primarily by way of court actions.

This last chapter focuses on the future. In the next few decades, the global economy will undergo profound paradigm changes, resulting from a variety of technological advances, including, as relevant here, the advent of 3D printing technology. Great expectations are pinned to this technology in terms of initial changes in the production process followed by changes in business models. 3D printing technology will undoubtedly find widespread application in the automotive industry, in particular the spare parts industry.[1] Of course, mass production of spare parts is still several years away, estimated in the range of 10–15 years. Once established however, the technology will profoundly change the mode of industrial production, as well as the market structure and the relative positions of its players.

This chapter will consider the potential impact on the spare parts market from an intellectual property perspective. It will start with the technical background of 3D printing followed by an overview of its impact on industry and economy. Next we will turn to the market for visible spare parts, again with the automotive industry as

[1]The technology is already being applied in the automotive parts industry. Entire cars have been produced by 3D printing. Manufacturers also use 3D printing on a routine basis to manufacture specific parts. Local Motors, a U.S. company founded in 2007, produced an entire car by using 3D printing, based on open-source and crowdsourcing (see https://localmotors.com/); to this, see Anastassacos 2015. The first 3D-printed car was *Urbee*, a prototype hybrid car whose body completely consists of Acrylonitrile butadiene styrene (ABS) plastic (see https://korecologic.com/). BMW is an example for an established OEM which already uses 3D printing to produce parts, even metal parts. For details on the development, see Sapienza 2015.

D. Beldiman and C. Blanke-Roeser, *An International Perspective on Design Protection of Visible Spare Parts*, SpringerBriefs in Law,
DOI 10.1007/978-3-319-54060-3_7

the leading example, and conclude with an assessment of the impact of 3D printing technology on the competitive advantage conferred by intellectual property, including design law and technological know-how.

7.2 Some Background on 3D Printing

Unlike traditional manufacturing methods, which produce a specific 3-dimensional shape by *removing* material (subtractive manufacturing), 3D printing is a method that operates by *adding* material layer by layer in order to achieve the desired shape (additive manufacturing).[2] Among other advantages, it allows creation of shapes, such as hollow spaces, which cannot effectively be achieved by conventional technologies, like casting, molding or machining. A broad variety of 3D printing methods and materials are available.[3] Theoretically, almost every material is suitable for 3D printing—from classical material such as plastic to biological cells.[4] Almost everything will be producible by 3D printing,[5] including, for present purposes, parts of complex products, such as cars.[6] Production in a 3D printing setting is controlled by a computer on the basis of a computer-aided design (CAD) program that contains digital instructions directed to the 3D printer.[7] This program generally resides in an electronic file (CAD file) associated with a particular product.[8]

3D printing technology has been in use for decades in some industrial environments. For instance, in the airplane industry, Boeing[9] and Airbus[10] routinely use specialized 3D printers for manufacturing certain parts that cannot be produced in other ways. With the development of cheaper and more generally applicable printers, the technology is spreading to private households as well.[11] Besides industrial and home use, professional 3D printing shops and online services are emerging.

Use of the 3D printing technology has accelerated in recent years as a result of several factors. These include the increasing potential of 3D printing technologies

---

[2]For a more detailed introduction, see Zukas and Zukas 2015, para 1.1. For details on the history of 3D printing, see Mendis et al. 2015, p. 2. Some scholars differentiate the terminology, using "additive manufacturing" for industrial settings and "3D printing" for media or hobbyist environments. Others use the terms synonymously. See e.g. Europe Economics 2015, p. 127; Huang et al. 2015. For purposes of this book the term 3D printing will be used throughout.

[3]For details on the different methods, see Bechtold 2016, p. 519.

[4]Cf. Bechtold 2016, p. 520.

[5]Bechtold 2016, p. 519.

[6]See also Bechtold 2016, p. 520.

[7]Nordemann et al. 2015, p. 1265.

[8]Nordemann et al. 2015, p. 1265; Schmoll et al. 2015, p. 1042: CAD files may either be custom developed or downloaded from the internet, e.g. special online platforms.

[9]See http://www.boeing.com/features/2016/08/record-books-08-16.page.

[10]See http://www.airbusgroup.com/int/en/story-overview/Innovators-3D-printing.html.

[11]For details, see Bechtold 2016, pp. 524 et seq.

and materials, their decreasing cost, and the expiration of patents on relevant 3D printing technologies and materials.[12] This has allowed a number of companies to enter the field, with the result that a fairly robust, if technologically incipient, 3D printing industry has emerged.[13] The 3D printing industry's annual growth rate is estimated at 60%,[14] but widespread industrial use is still several years ahead. In the longer term future, 3D printing is expected to operate a fundamental paradigm shift in the existing business models of the manufacturing sector.[15]

7.3 Impact of 3D Printing Technology on Manufacturing Business Models

3D printing technology arrives at a time when business models are undergoing a profound shift, associated with global, internet-enabled technologies and modes of collaboration. Organizational science refers to this as the "Big Shift", the move away from models based on a "stock" of assets, to a model of "flow", characterized by the ability to capture and employ flows of knowledge. In this model, competitive advantage comes from rapid deployment of new technologies, flexibility of production, connection to networks, decentralization and proximity to customers.[16] 3D printing is one of the main drivers of the Big Shift. It is likely to disrupt virtually every aspect of production, from initial design to manufacturing and assembly, to supply chain, logistics, distribution, and consumption, with a global effect.[17]

Scholarly literature describes the difference between conventional business models and a 3D printing environment as follows.[18] The conventional model of industrial production is based on economic scale and accumulation of stocks of proprietary assets, including knowledge, capital and other resources. Competitive advantage is gained using such accumulated assets towards efficient management of production and distribution in an economy of scale.[19] In an economy of scale, up-front tooling costs are considerable and must be amortized. This requires production capacity to be consolidated and centralized. While initial setup costs are high, the incremental cost to produce each part is low. This drives production of as

[12]Cf. Markets and Markets 2015, paras 7.3, 7.5.

[13]Smaller players such as 3D Systems and Stratasys have gained in volume, but large players, in particular Hewlett Packard have begun acquiring 3D printing technology for the industrial sector. See http://fortune.com/2016/05/17/hp-3d-printer-manufacturing/.

[14]Cohen et al. 2014.

[15]Adler 2014, pp. 24 et seq.

[16]Hagel III et al. 2009, pp. 1 et seq.; see also Birtchnell and Urry 2013, pp. 33.

[17]Bechtold 2016, p. 518 et seq.; Kylau et al. 2015; Lemley 2015, pp. 461 et seq. For an overview, see Campbell et al. 2011, pp. 10 et seq.

[18]Hagel III et al. 2009, pp. 1 et seq., who titled it "The Big Shift".

[19]Hagel III et al. 2009, pp. 1 et seq.

many parts as possible.[20] Large volumes of products, in turn, demand widespread distribution networks and a massive marketing and sales apparatus.[21] Competitive advantage lies in scale.[22]

In a 3D printing environment, these constraints are expected to largely disappear.[23] With decreasing cost of 3D printers, little up-front tooling is required. In many cases, the same equipment can be used to manufacture different types of parts. Therefore, relatively short setup time is required and the overall timeline from initial design to prototype to finished product is shortened. As a result, the need to amortize setup costs over large production runs is deemphasized. When manufacturing and assembly are decentralized and located in greater proximity to customers, the need for shipping will decrease, as only raw materials and the design file need to be shipped, bypassing traditional manufacturing, supplier, and distribution channels altogether.[24] Raw materials can be delivered for just-in-time fabrication to a local service hub where 3D-printed components are installed into a complex product, at the time of use.[25] With that, the distribution costs would drop significantly.[26] Storage costs decrease, largely as a result of simplified equipment and increased efficiency in producing on demand.[27] The proximity to customers will allow for a closer interaction between distributors and customers. Reduction in production and distribution costs may make even low-volumes production economically attractive.[28] This would lower barriers to market entry, opening the doors to new players and result in decentralization.[29] As a result, new production and

[20]Kylau et al. 2015.

[21]Cf. Widmer and Rajan 2016, p. 13.

[22]Cf. Widmer and Rajan 2016, p. 13.

[23]Cf. Widmer and Rajan 2016, p. 3; Reeves and Mendis 2015, p. 14; Giffi et al. 2014, p. 5.

[24]Cf. Reeves and Mendis 2015, p. 14: "compressing the supply chain"; Campbell et al. 2011, p. 11: "Manufacturing could be pulled away from 'manufacturing platforms' (…) and back to the countries where the products are consumed, from the home to other larger but local facilities"; more skeptical White 2015.

[25]See Birtchnell and Urry 2013, p. 33.

[26]Cf. Giffi et al. 2014, p. 4; Lemley and Shapiro 2013, pp. 1135 et seq.; Campbell et al. 2011, p. 10. See also the discussion of outsourcing the production to professional printing shops close to end customers in Bechtold 2016, p. 531.

[27]White 2015; see also Lemley and Shapiro 2013, pp. 1135 et seq.; Campbell et al. 2011, p. 10.

[28]See Lemley and Shapiro 2013, pp. 1135 et seq.

[29]Cf. Reeves and Mendis 2015, p. 14: For example, this may incentivize new manufacturers of spare parts for more exclusive cars, whose businesses are based on small-scale production; see also Birtchnell and Urry 2013, p. 27. Early stages of this shift have already occurred. Few would have imagined 20 years ago that the distribution systems of established industries, such as record labels and the movie industry, could be fundamentally disrupted by the internet. Or even five years ago, that the hotel industry could be cannibalized, by businesses relying not on accumulated assets of hotel space, but on the ability to use platforms that capture and employ flows of knowledge about available space among private parties (e.g. Airbnb). See Widmer and Rajan 2016, p. 4; also Lemley and Shapiro 2013, pp. 1135 et seq.; Surden 2013, pp. 1 et seq.

distribution channels, such as e-tailing, could develop.[30] In turn, these changes will produce a shift in the source of competitive advantage from large scale production towards other areas of the value chain, such as speed to market, flexibility, IP ownership of platforms, ownership of customer networks and collaborative customer relations.[31]

7.4 3D Printing in the Automotive Aftermarket

In the automotive spare parts industry, 3D printing is already being applied on a limited basis,[32] but once more widely spread, it is expected to bring about profound changes in the existing business models.[33] These include, as relevant for present purposes, changes in the structure of the automotive spare parts production, a shift in the source of competitive advantage from scale economy to information assets and the ease of creating customized new lines of products that better meet consumer preferences.[34] Nonetheless, it must also be kept in mind that massive use of 3D printing technology for automotive spare parts remains a matter of the future and, depending on the course of technological development, the expectations may not be met in their entirety in some areas. We will next discuss some of these expectations and concerns.

7.4.1 Changes in the Market Structure

The spare parts industry may be among the first sectors to undergo change as a result of this new technology.[35] One of the early developments will be entry into the market of a diversity of new players. Because no up-front tooling and little setup time are required, the overall cost of bringing new products to market will decrease. This in turn will lower barriers to entry and allow new competitors to access the

[30]E-tailing means selling retail goods online. On the future development, see e.g. Deutsche Post DHL, Global E-tailing 2025, 2014, available at: http://www.dpdhl.com/content/dam/dpdhl/global-etailing-2025_de.html.

[31]See Lemley and Shapiro 2013, pp. 1135 et seq.

[32]See also Frost and Sullivan 2015, para 8: In 2015, 3D printing in the automobile industry was still mainly (90%) used for prototyping. However, even complete cars have already been produced.

[33]See Depoorter 2014, pp. 1485 et seq.

[34]See Anastassacos 2015.

[35]See Anastassacos 2015; Frost and Sullivan 2015, para 8.

market. Many of these will be smaller companies with more nimble business models that compete based on speed and closeness to the consumer.[36]

On the other hand, the complexity of automotive products requires investment in large-size professional printers, expensive materials and a fairly high level of expertise.[37] This means that companies performing 3D printing in the spare parts aftermarket will be professional services of a certain size. Home production on any meaningful scale is not expected.[38] It is likely that the numerous local entities[39] that will emerge, will include both OEM-authorized and independent suppliers or repair shops.[40]

3D printing requires a high degree of specialization in the individual operations involved: CAD design, 3D printer manufacture, operation, and maintenance; raw material preparation and management; analysis of finishing; and supply chain and project management. With the expansion of 3D printing, a greater need for formal and extensive training and skill development programs will emerge. It is conceivable that talent shortage could constitute a bottleneck in the expansion of the technology.[41]

The decentralized locus of production and the increased number of players in the market may give rise to further collaborative relations. If localized spare parts production is possible at the requisite qualitative level and demand exists, manufacture of selected parts might be licensed to qualified local entities, providing them with the requisite CAD files and raw materials. Additionally, customers might take a more active part in the configuration and design of the product, by choosing products online, modifying their design ("personalized products") and having them printed by local services. These could also provide assistance and the advice relating to the choices available.[42] Licensing the manufacture of parts to numerous

[36]See Birtchnell and Urry 2013, p. 27.

[37]For example, an average industrial 3D printer costs about $75,000. See Europe Economics 2015, p. 130.

[38]To this, see Europe Economics 2015, pp. 131 et seq., 160 et seq.

[39]A higher price pressure on OEMs can be expected, see Anastassacos 2015.

[40]Moreover, the costs for 3D printing are cheaper for professional users than they are for private consumers, as they are producing on a large scale and employing specialists (Europe Economics 2015, pp. 128 et seq., also for details on the costs of 3D printing). See also Dawson 2013, citing *Sandro Piroddi*, Supervisor for Rapid Prototyping, Ford: "Piroddi does not expect consumers to be producing and fitting their own spare parts at home either. 'You will not reach the quality we have here, or on the expensive printers (…) I can imagine if the printer is a low price, you have to pay a high price for your material. The next thing is you need a certain software to design your part, to treat it so that you can build your part on your printer. But perhaps, you can get some files of parts which are already constructed from the internet, and print them at home (…) On the other hand, never say never – I don't know what will happen in 100 years'".

[41]Giffi et al. 2014, p. 5.

[42]See also Europe Economics 2015, p. 130.

smaller service providers could develop into an attractive business model, of which both OEMs and independent spare parts manufacturers might want to take advantage.[43]

7.4.2 Changes in the Source of Competitive Advantage

If physical plants, production lines and large-scale production cease to provide competitive advantage in the market, the question arises what will replace them. It is commonly accepted that in a knowledge economy, information in the form of technical know-how and expertise, production know-how, understanding and knowledge of the customers, etc. will constitute the differentiating assets among market participants.[44] Ownership of or access to the relevant information assets will, therefore, be the source of competitive advantage to any player in the market.[45]

From an IP law perspective, the information assets primarily relevant to the production of visible spare parts are a part's design and the associated technological know-how.[46] Design represents a product's outer appearance, which is usually one of its important selling points. The customization and personalization opportunities afforded by 3D printing technologies are likely to become attractive revenue generators, as discussed in further detail below.[47] Design can be appropriated and protected under industrial design laws. However it is subject to "leakage" in the sense that once disclosed, it is easily replicable. Its infringement, given numerous players and widely distributed locations of the infringement in a 3D printing environment, will be increasingly difficult to enforce.

Technological know-how has always been important in the automotive industry, but with the advent of "intelligent" vehicles, the technological sophistication of modern automobiles will reach an entirely new level. Visible spare parts will be implicated in these changes. For instance, cars with progressively more autonomous maneuvering features, such as self-parking, hands-free and, ultimately, fully

[43]For customers, it might be interesting to choose the original product if there is no vast difference in prices, as parts produced on the basis of the original CAD files would often be seen as more trustworthy, standing for higher quality and safety standards. Negotiation on the amount of royalties will become critical. As independent spare parts manufacturers would also be interested in such business models, royalties would become one decisive factor determining market positions.

[44]See generally Hagel III et al. 2009, pp. 1 et seq.

[45]Other elements of the competitive advantage include speed of market launch of products, flexibility, IP rights, customer networks and collaborative customer relations and branding. See Lemley and Shapiro 2013, pp. 1135 et seq.

[46]Technological knowledge may in principle be patentable, but often in spare parts does not meet the criteria for patentability. The following references to technological know-how relate to unpatented, confidential information, that would likely be protected under trade secret laws.

[47]See infra section 7.4.3.

autonomous driving, must accommodate an increasing number of sensors, cameras and radars in their outer shells. The shape of body parts impacts the placement and operation of these devices and utmost precision is required in the production and placement of the parts that accommodate them. Furthermore, in electric vehicles, for instance, body parts themselves may contain technology that performs further functions: trunk lids, door panels and hoods can be manufactured from electrically chargeable materials, so that these parts effectively act as batteries.[48] It is expected that at some point in the future, many of these highly complex parts can be produced by 3D printing technologies.

Production of such parts requires technological know-how, not only relating to traditional automotive safety, such as crashworthiness, but also relating to sensors, transceivers and yet to be developed technology relating to the vehicle's "intelligence" as well as their interoperation with the vehicle's computer system. This technical information is captured in software developed for use on 3D printers and includes the design's CAD file, as well as the relevant technical specifications for materials, manufacture and installation.[49] The data are proprietary to the automotive manufacturer and are normally not disclosed to unauthorized third parties. Without them, it will be virtually impossible to produce marketable parts for an advanced vehicle.[50] Access to such proprietary data has already become the subject of disputes between OEMs and independent suppliers in various areas of automotive services, and with increasing complexity of the vehicles, is likely to sharpen.

7.4.3 Customization

A further trend facilitated by 3D printing is design customization. The low cost and accessibility of 3D printing technologies encourages experimentation with the customized appearance of products. It will fuel new demand by consumers,[51] who view the visual differentiation and distinctiveness of products they own, as an expression of identity.[52]

---

[48]Volvo announced a feature of its electric vehicles: electrically chargeable body parts composed of multiple layers of carbon fiber, which are insulated from each other by fiberglass inserts. The layers of polymer-infused carbon fiber are actually acting as the cathode and anode in this system with super capacitors built into the skin. See Whitwam 2013.

[49]Giffi et al. 2014, p. 12; Walter-Hermann and Büching 2014.

[50]In the context of intelligent vehicles, mere reverse engineering rarely yields sufficient information to support production of a part.

[51]For the special case of the so-called *FabLabs* (fabrication laboratories) – open, democratic workshops where users can produce things using technologies such as 3D printing, connected with models like open-source hardware, see Anastassacos 2015 (for implications on the spare parts market); also Walter-Hermann and Büching 2014.

[52]See generally Beldiman 2015, pp. 104 et seq.

This trend is not entirely new. Customization of vehicles is already practiced to some extent. OEMs offer a selection of different designs of wheels, grilles, lights, etc. Some of these parts can be used for many different models, following a modular system.[53] The freeform capabilities of 3D printing will allow for individualized, personalized designs.[54] Authorized customization shops[55] are offered by OEMs and independent customization services abound.[56] Manufacturers will be quick to capitalize on these new business opportunities. It is important that customized parts be within approved parameters, e.g. NHTSA crash performance norms. Therefore, production must be preformed by specialized entities. Future customers may to some extent participate in the personalization process, even shaping their own models as part of the order process, ("design yourself").[57] However, their input is likely confined to the aesthetic and functional limits set by the manufacturer.[58]

To what extent customization may also occur at the level of the IAM is uncertain. Pressure from customers in this direction will undoubtedly exist. While 3D printing will largely overcome technical limitations, it will not completely eliminate them. Moreover, with growing technological complexity and stringent regulatory norms,[59] it is uncertain whether independent service providers could produce more than very basic customized exterior parts, without access to the requisite technical information.

7.4.4 Limiting Factors

To place the foregoing in proper perspective, it must be kept in mind that large-scale production of automotive spare parts by 3D printing remains a thing of the future, which has had only limited impact on the market so far. It may take some 10–15 years for this to change.[60]

---

[53]See Reeves and Mendis 2015, pp. 14. For details on this principle and case studies from the French automobile industry, see Doran et al. 2007, p. 2 et seq.

[54]An interesting segment of the automotive industry that has already adopted 3D printing is the "ultraluxury" segment. See Giffi et al. 2014, p. 11.

[55]Examples are: AMG, Brabus and Abarth.

[56]See e.g. http://www.autoanything.com.

[57]Cf. Reeves and Mendis 2015, p. 20, who state that this will rather be a "long-term opportunity". Likely this process will require assistance from qualified employees. Depending on the amount of freedom in designing such parts, designs may depart from the standard to an extent that they could become the customer's own design.

[58]Ultimately customers select automobiles for the aesthetic appeal imparted to them by professional automotive designers, carefully harmonized in terms of aesthetics and functionality. Outside input may be perceived as undesirable by the manufacturer and ultimately not result in an appearance pleasing to the customer.

[59]Some OEMs offer numerous alternatives within approved parameters, e.g. NHTA crash performance norms. Therefore, replication must be done by specialized entities.

[60]See also Mendis et al. 2015, p. 6: ten years; Reeves and Mendis 2015, p. 14: 15 years.

Some technical questions remain open. Currently, industrial 3D printers and materials are still quite expensive. While in the course of the years their costs will drop and their quality potential will rise, it is uncertain whether they are capable of producing automobile parts of the same quality standard as the ones manufactured conventionally.[61] Technical limitations of 3D printing technologies, at least at its current level, may lead to differences in mechanical behavior, strength or durability of the material in the goods produced by this method,[62] compared to their conventionally produced counterparts.[63] Three concerns are commonly articulated in this regard: that the materials used in 3D printing are weaker than those used in conventional production; that different printers can lead to different stability standards despite the use of the same CAD file (lack of "process repeatability") and that the layer-by-layer structure leads to uneven stability of the product.[64] The last item is of greatest concern, as it may constitute a structural defect of the 3D printing technology in general. In the automobile industry, where quality and safety are key, differences in material properties are critical. By way of example, metallic parts, such as body parts, whose stability is a function of material density, are produced by the process of sheet metal stamping.[65] Experts are debating whether 3D printed products will reach the quality of their conventionally manufactured counterparts.[66] On the other hand, in the future, the structure and main material of cars could change altogether, leading to different mechanical properties, especially with regards to the transfer of impact energy in case of a crash. Furthermore, other reasons may prevent parts from being produced by 3D printing, including the fact that technological qualities of 3D printers are not suitable for certain parts[67] and that some car parts may be simply too big to be printed by 3D printers that fit into workshops or shops.[68]

Other factors limiting the use of 3D printing include the possible lack of CAD files for many spare parts: for parts that OEMs do not produce by 3D printing technology, no "official" CAD files will exist.[69] In order to produce the part,

---

[61]Mendis et al. 2015, p. 6; see also Campbell et al. 2011, p. 7.

[62]Reeves and Mendis 2015, p. 19.

[63]Reeves and Mendis 2015, p. 19.

[64]Campbell et al. 2011, p. 7. To this and additional technological limitations of 3D printing, see Ford 2014, p. 23; Bourell et al. 2009, pp. 16 et seq.

[65]Reeves and Mendis 2015, p. 19.

[66]According to Reeves and Mendis 2015, p. 18, some parts will never be produced in the same quality as it is possible by means of traditional manufacturing. Rather optimistic Giffi et al. 2014, p. 13, providing a graphic that shows which parts could be manufactured by means of 3D printing in the future. On the other hand, 3D printed parts are already being used in the airplane industry (see infra section 7.2).

[67]Reeves and Mendis 2015, pp. 16 et seq., also with a list of example parts; skeptical also White 2015. In contrast, according to Markets and Markets 2015, para 6.3.2, components such as the frame and doors will be producible by 3D printing one day.

[68]Cf. Reeves and Mendis 2015, p. 18; see also Giffi et al. 2014, p. 17.

[69]Reeves and Mendis 2015, p. 19.

independent suppliers will have to create their own files, mainly by 3D scanning existing original parts. Even with an intact specimen, it is hardly possible to create a CAD file that contains all relevant information on the original part's properties, including the safety-relevant properties by using a scanner.[70]

Finally, there is the question of cost. Even though a part may be technically produceable by means of 3D printing at the same quality as by conventional manufacture, 3D printing will be the more expensive choice, as mass production would be cheaper.[71] This disadvantage will not necessarily be outweighed by the above-mentioned advantages of 3D printing.[72] Therefore, 3D printing will be mainly relevant for parts where customers wish individualization.[73] Whether future innovation will overcome these shortcomings is uncertain, as is the ultimate future role of 3D printing. It may not completely replace conventional manufacture of spare parts as the only manufacturing technique in the future, but will undoubtedly play an important role in shaping the global automotive landscape.[74]

7.5 The Legal Debate on Visible Spare Parts in a 3D Printing Environment

Having examined the impact of 3D printing technologies from a business perspective, the question arises what effect, if any, 3D printing will have on the legal debate surrounding visible spare parts. As a general proposition, assuming the current configuration of interests persists, in which OEM and IAM markets compete for larger market shares, it is fairly certain that the debate over spare parts will

[70] Reeves and Mendis 2015, p. 19. On the limits of 3D scanning Allen 2013.

[71] In future, it can be expected that 3D printing technologies will become faster, raising their potential for mass production in the industry (cf. Kelly Services 2016: "(T)hough 3D printing isn't yet feasible for use in mass production, it's probable that in the near future, it will be"). However, this is not the question here, as this section deals with the use of 3D printing technologies at a local level, producing single parts on demand only. Furthermore, next to the above-mentioned costs for printing technologies and materials, there will be labor costs remaining, caused by the necessary pre- and post-printing processing, e.g. polishing and cleaning the product; see Holweg 2015.

[72] Reeves and Mendis 2015, pp. 16 et seq., also with a list of parts; see also Allen 2013.

[73] Holweg 2015: "3D printing simply works best in areas where customization is key (…) However, we also know that 99% of all manufactured parts are standard and do not require customization".

[74] See also Holweg 2015; Park 2015: "(I)t can not now, and likely will not ever exist in isolation"; Giffi et al. 2014, p. 19: "While (3D printing) will not become the only manufacturing technique in the future, it will nonetheless play an important role in shaping the global automotive landscape"; Timms 2014. For an interesting analysis of factors that will influence the role of 3D printing in future economies, see Birtchnell and Urry 2016, pp. 88 et seq.; Reeves and Mendis 2015, p. 18; Wong and Hernandez 2012, p. 8. Indeed, 3D printing plays a relatively greater role in the "ultraluxury" segment of the automotive industry already today.

continue and even sharpen. As for more detailed aspects, several points are discernable, with the proviso that foretelling the future is inherently uncertain.

First, the need for design protection is likely to increase, while its enforceability will decrease. As discussed above, the ease of customization facilitated by 3D printing technology, combined with customer demand for new products, will present new business opportunities.[75] In that scenario, the customized appearance of the outer shell of a car is likely to become a selling point.

Spare part designs may originate from different sources, including the OEM market, the IAM or the customer. OEMs will continue to secure protection of the spare part designs they create. Over time, current practices will change to add a growing number of alternative designs for each spare part. Protectable designs may also originate in the IAM. Independent service providers may either design and build complete parts on their own (assuming that is possible, given the technical limitations), or merely design the appearance and license the technology from an OEM, although it is not clear that OEMs would necessarily be interested in such deals. Further, the scope of the design freedom of an independently designed part destined for installation on an OEM vehicle may be restricted. OEMs might object to IAM designed parts, on the ground that their unauthorized installation would alter the overall appearance of the OEM's vehicle's design. This would open up a new battleground between the stakeholders in the spare parts debate and would increase reliance on protecting the design of the entire vehicle. Finally, designs may be generated with customer participation. How the customer's creative contribution is to be handled from a proprietary perspective is a question that remains to be resolved. The solution may well be an assignment of rights to the producer of the part, assuming that the rights are in fact fully assignable.[76]

Thus the need for design protection for purposes of product differentiation may be growing. However, its enforcement against unauthorized 3D printed spare parts, in the context of widely distributed local production sites, is virtually impossible. Therefore, the extent to which design protection confers actual competitive advantage is not clear, other than possibly through the deterrent effect of registration.

Second, the role of the repair clause as a challenge to design protection will diminish. Independent suppliers will continue trying to challenge OEM design rights to spare parts in various ways, including validity,[77] exhaustion and the repair clause. The repair clause itself however, is likely to weaken as a defense, because its

---

[75]The contrary argument is that such aesthetic differentiation usually comes at the end of a technological cycle, when functional improvements can no longer be made, e.g. different colors of notebooks. The automotive industry is entering a new age of assisted and automated driving. It might be possible for these developments to overshadow any aesthetically based alternate products offered by manufacturers.

[76]This would be a question for the national law of individual jurisdictions.

[77]Another challenge to spare part design rights that might be impacted in a 3D printing environment is the functionality defense, as customization of a vehicle's spare part's design is likely to be based on aesthetic considerations.

conceptual basis may well erode in an environment of substantial customization of a vehicle's exterior. As currently enacted in some countries, the applicability of the repair clause operates as a defense to design infringement, where a spare part is used for repairs to restore a product "to its original appearance". The repair clause is meant to prevent the design holder's monopoly over spare parts that "must match" the original in order to allow restoration of the vehicle's original appearance.[78] If a part "must match" the original, which in turn is design protected, the design owner holds a full monopoly for that part and no opportunity for substitution by a competitor exists. By exempting such parts from design protection, the repair clause gives competitors a chance to supply the spare parts market.

According to recent jurisprudence, the clause applies only as long as there is no substitutability, i.e. where the parts are "dependent" on the appearance of the original vehicle,[79] or, alternatively stated, where an "objective necessity" that they match the original vehicle exists.[80] In an environment of increasing customization, in which the manufacturer offers multiple lines of parts for the customer to choose, the concept of "the" original appearance of the vehicle is relativized. If consumers can select one among multiple lines of wheel caps, or alternatively, different designs of grilles, lights, windows, etc. these parts are discretionary to the appearance and no longer capable of creating a monopoly for the design holder. With this, the main argument in favor of the repair clause would fall. This trend has already commenced.[81] As the degree of customization increases, the number of discretionary parts will grow as well, causing the role played by the repair clause to shrink.

By the same token this will open up opportunities for the aftermarket. If the appearance of vehicles is no longer strictly tied to the specific appearance of a specific model, independent tuning and modification services may offer to upgrade vehicles based on their own designs. These parts can range from wheel caps to entire body kits. This means that competitive opportunities would no longer be foreclosed by the design protection of a part that "must match" a vehicle in order to restore its original appearance.[82]

Third, the true competitive advantage resides in technological know-how. As indicated above, design rights may be increasingly hard to enforce. Reliance on

---

[78] Article 110(1) CDR.

[79] BMW AG *v.* Round and Metal Ltd and Philip David Gross [2012] EWHC 2099 (pat).

[80] BGH, 2 June 2016, I ZR 226/14 (OLG Stuttgart), GRUR Int 2016, 1057 – Kraftfahrzeugfelgen; OLG Stuttgart, 11 September 2014, 2 U 46/14, GRUR 2015, 380.

[81] See the above-mentioned recent referrals to the CJEU: Corte di Appello di Milano, 18 July 2016, Case C-397/16 – Acacia s.r.l. v Pneusgarda s.r.l. (in bankruptcy), Audi AG; BGH, 2 June 2016, I ZR 226/14 (OLG Stuttgart), GRUR Int 2016, 1057 – Kraftfahrzeugfelgen. For details on them, see supra section 5.2.3.

[82] Aftermarket tuning services have long been available offering entire body kits or replacements of body panels, bumpers, side skirts, fender flares, spoilers and hoods. However, their scope and flexibility is likely to be considerably enhanced in a 3D printing environment. For examples of aftermarket tuners, see Teslarati Network, Tesla Model S Aftermarket Body Kits, available at: http://www.teslarati.com/tesla-model-s-aftermarket-body-kits/; http://www.revozport.com/tesla/quality-control.html.

design protection alone thus may be insufficient.[83] Know-how on the other hand, is subject to dual protection: from a practical perspective by way of secrecy and from a legal perspective by way of trade secret law.[84] Know-how is a powerful competitive differentiator: without access to the OEM's know-how, a competitor is likely precluded from replicating a visible spare part, as in intelligent vehicles reverse engineering alone does not provide the requisite information.[85] On the other hand, maintaining secret technological know-how is not without its challenges. The know-how captured in software programs, in effect becomes a digital good, subject to all the risks of digital distribution, including hacking and piracy. Once the information has been misappropriated and spread, legal remedies are largely inadequate, in particular, in a setting of globally distributed manufacturing capabilities.

Overall, 3D printing may not clearly favor either OEMs or the IAM. Instead its effect may well be a relaxation of the conflict between stakeholders and a more collaborative approach in taking advantage of the opportunities this new technology has to offer. As regards design protection, customization and the ease and low cost of production may give rise to such a multitude of new design options for both OEMs and independents, that the tension relating to specific designs (a scarcity of designs) will be defused. Technological know-how, on the other hand, will remain "scarce" because it is protected and will confer an advantage to its owner. It is conceivable that this advantage would be perceived as an "information inequality" sufficient for regulation to intervene in some form.[86] If so, the most likely means would be competition law, although success of such efforts is questionable, given European Commission rulings that have declined to find competition violations in the automotive spare parts market.[87] It is more likely that OEM's will take advantage of the numerous smaller local players to enter cooperative arrangements that allow localized, on demand supply of spare parts.

---

[83]Exterior spare parts are traditionally hard to protect by patents. However, that may change with intelligent vehicles.

[84]Directive (EU) 2016/943 of the European Parliament and of the Council of 8 June 2016 on the protection of undisclosed know-how and business information (trade secrets) against their unlawful acquisition, use and disclosure, L 157/1; U.S. Defend Trade Secrets Act of 2016.

[85]Simple reverse engineering alone of a visible spare part does not ensure its interoperability with the vehicle's electronics.

[86]A possible basis for challenging an IP owner's market dominance is competition law, in the event it has an adverse impact on consumers. It might be argued that the OEM's technical information constitutes an essential facility and excludes competitors. See CJEU, 29 April 2004, Case C-418/01 – IMS Health GmbH & Co. OHG v. NDC Health GmbH & Co. KG [2004] I-5039; CJEU, 6 April 1995, Joined cases C-241/91 P and C-242/91 P – Radio Telefis Eireann (RTE) and Independent Television Publications Ltd (ITP) v. Commission of the European Communities (Magill) [1995] I-743. Compulsory license solution actions challenging OEMs' failure to release proprietary know-how on competition grounds have been brought in various fora.

[87]CJEU, 5 October 1988, Case C-238/87 – AB Volvo v. Erik Veng (UK) Ltd. [1988] 6211; CJEU, 21 June 1988, Case C-53/87, – Consorzio Italiano della Componentistica di Ricambio per Autoveilici (CIRCA) and Maxicar v. Régie Nationale des Usines Renault [1988] ECR 6039, 237.

References

Adler R (2014) Navigating Continual Disruption, A Report on the 2014 Aspen Institute Roundtable on Institutional Innovation. http://www.johnseelybrown.com/navigatingdisruption.pdf

Allen N (2013) Why 3D printing is overhyped (I Should Know, I Do it For a Living). http://gizmodo.com/why-3d-printing-is-overhyped-i-should-know-i-do-it-fo-508176750

Anastassacos T (2015) 3D printing will disrupt the spare parts market. The question is how, how soon and what to do about it, Service in Industry Briefing. http://serviceinindustry.com/2015/03/03/3d-printing-will-disrupt-the-spare-parts-markets/

Bechtold S (2016) 3D printing, intellectual property and innovation policy. IIC 47:517–536

Beldiman D (2015) Expressive dimensions in design: a question of incentive? In: Beldiman Dana (ed) Innovation, competition, collaboration. Edward Elgar Publishing, Cheltenham (UK), Northampton, MA (USA), pp 104–131

Birtchnell T, Urry J (2013) 3D, SF and the future. Futures 50:25–34

Birtchnell T, Urry J (2016) A New Industrial Future? 3D printing and the reconfiguring of production, distribution, and consumption. Routledge, London, New York

Bourell D, Rosen DW, Leu MC (2009) The roadmap for additive manufacturing and its impact. Identifying the future of freeform processing. https://wohlersassociates.com/roadmap2009A.pdf

Campbell T, Williams C, Ivanova O, Garrett B (2011) Could 3D printing change the world? Technologies, potential, and implications of additive manufacturing. Atlantic Council, Strategic Foresight Initiative. http://www.manufacturingglobal.com/technology/408/INDUSTRY-ANALYSIS:-The-pros-and-cons-of-3d-printing

Cohen D, Sargeant M, Somers K (2014) 3-D printing takes shape. http://www.mckinsey.com/business-functions/operations/our-insights/3-d-printing-takes-shape

Dawson R (2013) Cars in 3D—the role of 3D printing in the auto industry. http://automotivemegatrends.com/cars-in-3d-the-role-of-3d-printing-in-the-auto-industry/

Depoorter B (2014) Intellectual property infringements and 3d printing: decentralized piracy. Hastings L J 65:1483–1504

Doran H, Hill A, Hwang KS, Jacob G (2007) Supply chain modularisation: cases from the French automobile industry. Int J Prod 106:2–11

Europe Economics, The Economic Review of Industrial Rights in Europe—Final Report, MARKT/2013/064//D2ST/OP, Jan 2015, commissioned by the European Commission). http://ec.europa.eu/growth/tools-databases/newsroom/cf/itemdetail.cfm?item_id=8844

Ford SLN (2014) Additive manufacturing technology: potential implications for U.S. manufacturing competitiveness. JICE 1-35

Frost & Sullivan (2015) Executive analysis of 3-D printing in the automotive industry. http://www.frost.com/sublib/display-report.do?id=NFA0-01-00-00-00&src=PR

Giffi CA, Gangula B, Illinda P (2014) 3D opportunity in the automotive industry. Additive manufacturing hits the road. Deloitte University Press. https://dupress.deloitte.com/content/dam/dup-us-en/articles/additive-manufacturing-3d-opportunity-in-automotive/DUP_707-3D-Opportunity-Auto-Industry_MASTER.pdf

Hagel III J, Brown JS, Davidson L (2009) The big shift. Why it matters, deloitte center for the edge, September 2009. http://www.johnseelybrown.com/bigshiftwhyitmatters.pdf

Holweg M (2015) The limits of 3D printing. Harv Bus Rev. https://hbr.org/2015/06/the-limits-of-3d-printing

Huang Y, Leu MC, Mazumder J, Donmez A (2015) Additive manufacturing: current state, future potential, gaps and needs, and recommendations. J Manuf Sci Eng 137:014001–014001

Kelly Services (2016) Why 3D technology requires an adapting workforce and why the automotive industry cares. http://www.kellyservices.com.sg/US/Business-Services/Industry-Verticals/Why-3D-Technology-Requires-an-Adaptive-Workforce-and-Why-the-Automotive-Industry-Cares/?hid=RC#.WBHBucnpUbt

Kylau U, Goerlich K, Mitchell R (2015) How 3D printing will disrupt manufacturing. Don't dismiss it as a hobbyist's obsession. Make it part of your business strategy. Digit. Mag. http://www.digitalistmag.com/executive-research/how-3d-printing-will-disrupt-manufacturing

Lemley MA (2015) IP in a world without scarcity. NYU L R 90:460–515

Lemley MA, Shapiro C (2013) A simple approach to setting reasonable royalties for standard essential patents. Berkeley Technol L J 28:1135–1166

Markets and Markets (2015) 3D printing automotive market by technology (SLA, SLS, FDM, EBM, LOM, 3DIP), materials (Metals, Polymers), application (Prototyping & Tooling, R&D, Manufacturing), and region—global trends and forecast to 2020. http://www.marketsandmarkets.com/Market-Reports/automotive-3d-printing-market-250218997.html

Mendis D, Secchi D, Reeves P (2015) A Legal and Empirical Study into the Intellectual Property Implications of 3D printing. Executive Summary. Research commissioned by the Intellectual Property Office, UK. https://www.gov.uk/government/uploads/system/uploads/attachment_data/file/549044/Legal-Empirical-_Executive-Summary.pdf

Nordemann JB, Rüberg M, Schaefer M (2015) 3D-Druck als Herausforderung für die Immaterialgüterrechte. NJW, 1265–1271

Park R (2015) How 3D printing has disrupted the automotive industry. http://www.disruptivemagazine.com/features/how-3d-printing-has-disrupted-automotive-industry

Reeves P, Mendis D (2015) The current status and impact of 3D printing within the industrial sector: an analysis of six case studies. Research commissioned by the Intellectual Property Office, UK. https://www.gov.uk/government/uploads/system/uploads/attachment_data/file/549044/Legal-Empirical-_Executive-Summary.pdf

Sapienza JD (2015) Can we really 3D print cars and car parts? http://www.cheatsheet.com/automobiles/how-3-d-printing-will-transform-the-way-we-think-about-cars.html/?a=viewall

Schmoll A, Ballestrem J, Hellenbrand J, Soppe M (2015) Dreidimensionales Drucken und die vier Dimensionen des Immaterialgüterrechts. Ein Überblick über Fragestellungen des Urheber, Design, Patent und Markenrechts beim 3D-Druck. GRUR, 1041–1050

Surden H (2013) Technological Cost as Law in Intellectual Property. Harvard J L & Tech 27: 135–202

Timms M (2014) 3D printing cannot completely replace traditional manufacturing, say experts. http://www.worldfinance.com/infrastructure-investment/3d-printing-cannot-completely-replace-traditional-manufacturing-say-experts

Walter-Hermann J, Büching C (2014) FabLabs: of machines, makers and inventors. Transcript Verlag, Bielefeld

White G (2015) Industry analysis: the pros and cons of 3D printing. http://www.manufacturingglobal.com/technology/408/INDUSTRY-ANALYSIS:-The-pros-and-cons-of-3d-printing

Whitwam R (2013) Volvo develops battery-infused carbon fiber body panels. https://www.extremetech.com/extreme/168964-volvo-develops-battery-infused-carbon-fiber-body-panels

Widmer M, Rajan V (2016) 3D opportunity for intellectual property risk. Additive manufacturing stakes its claim. https://dupress.deloitte.com/content/dam/dup-us-en/articles/3d-printing-intellectual-property-risks/ER_2981-3D-opportunity-intellectual-property_MASTER.pdf

Wong KV, Hernandez A (2012) A review of additive manufacturing. ISRN Mech Eng 1–12

Zukas V, Zukas JA (2015) An introduction to 3D printing. Sarasota, FL (USA): First edition design publishing

Chapter 8
Conclusion

In the quest of a suitable treatment of spare parts under design law, the journey in this book has covered different countries and legal systems. While a solution has not emerged, several insights have.

It is clear that the automotive spare parts industry is a significant part of national economies worldwide. Its main stakeholders—the OEMs and independent suppliers—vigorously compete for market share, and as a result, are deeply divided on how the market for spare parts should be allocated. On the policy side, lawmakers are also split. In many countries, lawmakers are faced with conflicting priorities: liberalizing the spare parts market for the benefit of consumers or supporting national automotive industries, with a view to job creation and strengthening of the national economies. These differing approaches to the liberalization policy, lead to divergent interests among nations and the inability to reach a comprehensive international solution.

Design law happened to become one of the main legal battlegrounds in this dispute. Legal scholars were able to develop a narrowly tailored legal solution in the form of the repair clause, designed to allow liberalization of the spare parts market without unnecessarily impacting other aspects of the business. The appropriateness of this solution has become the topic of a doctrinal debate. Because the debate parallels the division of the stakeholders' economic and political interests it has resulted in an impasse. Consequently, on neither the doctrinal nor the legislative front, is there a prospect of imminent resolution imminently.

Against this background, a negotiated compromise may point the way forward. In theory, compromise should be feasible. Numerous aspects of the spare parts business, favor one or the other stakeholder, to the detriment of the other side, a circumstance that could lend itself to a trading of benefits. The bigger question is in what setting a negotiation could take place and, most importantly, what authority would have the necessary leverage to "strong arm" the players into concessions at a multinational level. The efforts undertaken by the EC have come closest in this regard, but so far without result. However, the issue may well be placed on the backburner, given new priorities on the policy front. If so, the legislative impasse in

D. Beldiman and C. Blanke-Roeser, *An International Perspective on Design Protection of Visible Spare Parts*, SpringerBriefs in Law,
DOI 10.1007/978-3-319-54060-3_8

the EU, as well as the U.S., may deter other nations from legislative action on this question, so that overall no legislative action is likely.

This leaves the courts as the battleground for the repair clause debate. It is possible that one of the numerous lawsuits filed throughout jurisdictions, may reach the highest court, and result in a dispositive precedential rule at national level, in favor of one or the other of the stakeholders, as did the Supreme Court in South Africa. It is impossible to predict what issues such cases will raise and the nature of a potential outcome. In the EU, for instance, movement in this direction may be triggered by the tension between a legally non-harmonized spare parts market and fundamental constitutional EU rights, such as the free movement of goods and services.

Still, the big picture that emerges is that design law on spare parts has essentially maintained a *status quo* for the past more than 20 years. Inevitably, this prompts a look at the future. Will the situation be the same 20 years from now?

Many take the view that change is imminent, albeit coming not from the legal, but rather from the technological side. New technologies, including most relevantly for present purposes, 3D printing technology are anticipated to fundamentally alter the business models in many industrial sectors. Manufacture of spare parts is expected to be one of the earliest industries impacted. With the decreasing cost of manufacturing in a 3D printing environment, manufacture of spare parts is expected to occur in a localized, "print-on-demand" fashion. New collaborative relations among players are expected to develop, including between OEMs and independent suppliers. The wide distribution of such services may lead to new types of licensing arrangements. Therefore, the battleground will likely move away from the manufacturing level, to the informational level. The outer shell of a vehicle is expected to play an increasingly important role in an era of intelligent vehicles and automated driving: new materials, sensors and communication devices incorporated into the external parts of a vehicle, make it part of a vehicle's "intelligence". The source of competitive advantage resides in the technological know-how, likely in electronic format. Separately, the new technology will allow increasing customization of the exterior of vehicles. The availability of multiple options for most must match parts, eliminates the ability to monopolize the design of a particular part. If that assumption is correct, the need for design protection will increase, however the need for a repair clause will diminish or disappear altogether.

Kolibri 360 GmbH
Ferdinand-Jühlke-Straße 7
99095 Erfurt, Deutschland
produktsicherheit@kolibri360.de